333.8
W 918 n

The Author

Robert E. Hunter is a senior fellow at the Overseas Development Council, Washington, D.C. and served as project director of the ODC'S recently published *The United States and the Developing World: Agenda for Action.* He is also professorial lecturer at the School of Advanced International Studies, Johns Hopkins University. Mr. Hunter received his Ph.D. from the London School of Economics, where he also taught. During 1964-65, he served on President Johnson's White House staff. During 1968-69, he was a research associate at the Institute for Strategic Studies in London.

The author wishes to express his appreciation to Mr. Malcolm Russell, School of Advanced International Studies, Johns Hopkins University, for his invaluable assistance in preparing this *HEADLINE SERIES.*

The Foreign Policy Association

The Foreign Policy Association is a private, nonprofit, nonpartisan educational organization. Its objective is to stimulate wider interest, greater understanding and more effective participation by American citizens in world affairs; and it presents this publication as a public service. However, as an organization, it takes no position on issues of United States foreign policy. In the *HEADLINE SERIES,* the author is responsible for factual accuracy and the views expressed.

HEADLINE SERIES, No. 216, June 1973, published February, April, June, October and December by the Foreign Policy Association, Inc., 345 E. 46th St., New York, N.Y. 10017. President, Samuel P. Hayes; Editor, Norman Jacobs; Associate Editor, Gwen Crowe. Subscription rates, $6.00 for 5 issues; $11.00 for 10 issues; $15.00 for 15 issues. Single copies, $1.25. Second-class postage paid at New York, N.Y. Copyright 1973 by Foreign Policy Association, Inc.

Library of Congress Catalog No. 73-80017

Energy Demand in the
United States

Since 1972, concern has been growing in the United States about an impending energy "crisis." This concern was dramatized by the shortage of fuel oil in the Eastern and Middle Western states during the winter of 1972-73; but the roots of the problem go much deeper, and its effects will be with us for a long time to come.

Few factors in American domestic and international life are so pervasive. Energy plays a crucial role in the lives of all Americans. It is vital to the highest standard of living on earth and to the style in which we live. It has been centrally involved in the debate during recent years over protection of the environment, whether from strip-mining, air pollution by the automobile or thermal pollution of lakes and streams. Energy has figured in discussions of tax policy because of the depletion allowance for investments in oil and natural gas; and its production problems have produced brownouts and black-outs in the United States during recent summers.

The international dimensions of energy supply and demand are much less well-known to most Americans, chiefly because we have not in the past been dependent upon other countries for supplies of the fuels that drive our economy. Of total U.S. energy consumption in 1972—a staggering 72,091 trillion BTU's, (or the equivalent of

12.7 billion barrels of oil)—only about 11 percent represented net imports from other countries. And most of this came from sources in the Western Hemisphere.

Traditionally, our energy supply has followed a pattern set for most segments of the economy. Of all the industrial nations of the West, we have been least dependent upon foreign trade for survival: we have depended on foreign sources for only a handful of strategic commodities; and only about 6 percent of our gross national product (GNP) is due to foreign trade. This situation has reflected the great wealth of our resources, and our ability and willingness to exploit them. In addition, it has reflected conscious decisions not to become economically dependent on the rest of the world. Of course, we have supported the expansion of international trade as a spur to our own economy; but we have steered clear of situations in which we might be "held to ransom" by others. Interestingly enough, the other superpower—the Soviet Union— long pursued a similar goal with even more consistency, while an incipient superpower—China—has pursued a policy of self-sufficiency with a particular passion.

Quite suddenly, however, it has dawned on the American public that the United States may not be able to be as self-sufficient economically as it has been in the past. The effects of inflation in recent years brought the first flashes of understanding, when the weakened position of the dollar led to its first—and then second— devaluation since 1933: the "impossible event." Now a so-called crisis in energy seems to be bringing new evidence to Americans that we will not be able to take economic shelter behind our ocean barriers in economic terms any more than we could do militarily, beginning a generation ago.

In one respect, the energy crisis is a problem of attitudes. We are having to alter long-standing patterns of thought and behavior, and most of us are finding ourselves unready for the adjustments that will be needed. Not surprisingly, therefore, much of the public discussion and debate on energy has had a naïve quality to it: it has been much like early writing on the nuclear arms race or on threats to the environment.

4

The role of American attitudes in defining a crisis can be illustrated by reference to attitudes elsewhere. For example, Western Europe imports more than 97 percent of its oil and Japan, its total consumption. Yet, despite increasing concern about certain facets of energy supply, in neither is there an energy crisis deriving from the mere fact of foreign dependence.

In addition, American psychological resistance to growing economic dependence on others—and especially energy dependence—is being exploited by some segments of the economy with a financial interest in the outcome of current debate or with a political purpose that can be advanced through economic nationalism.

In the debate on energy, it is also important to be cautious about relying on available statistics, either for demand or supply. Many of these statistics either rest on assumptions that could change, or are subject to challenge, especially in the case of proved reserves. Whether or not we really have an energy problem is thus partly a matter of guesswork; and predictions made in this study will be no better than the statistics now available. Indeed, a first requirement for the United States is to develop a far better set of publicly available statistics than heretofore, including oil company figures.

Because it will have an impact on so many facets of political and economic life—and because new facts and events are emerging every day—the energy problem cannot be dealt with between the covers of a single publication of this length. This study, therefore, will focus primarily on the problem's principal international dimensions: supply, security and cost. But to do that, it will be necessary first to look briefly at the domestic aspects of energy demand and supply. Hopefully, this *HEADLINE SERIES* will contribute to debate leading to a comprehensive national energy policy. In turn, such an energy policy could, with timely action, make the energy crisis a "self-denying prophecy."

'Insatiable' Demand

To any casual observer, U.S. demand for energy in all its forms—one-third of the world's total—must seem insatiable. We are the world's largest consumer of oil, natural gas and coal (although the

Table 1

World Energy Consumption—per capita (1970)

Country or Region	Coal Equivalent in kilograms
World	1,897
Market economies:	
Developed[1]	5,914
Other	330
Communist areas:	
Europe (including U.S.S.R.)	4,404
Asia	543
United States	11,128
Canada	8,997
Australia	5,374
United Kingdom	5,358
West Germany	5,151
U.S.S.R.	4,436
France	3,799
Japan	3,215
Italy	2,685
Argentina	1,686
China	522
India	189
Nigeria	45
Yemen	13

[1]North America, Western Europe, Japan, Australia/New Zealand, South Africa, Israel.
Source: *Petroleum Press Service*, April 1973.

Soviet Union is close behind in coal). We use rising amounts of nuclear energy and are passing Britain for the lead. Even in the relatively insignificant area of water-generated energy (4.1 percent of U.S. energy consumption) we lead all other countries in absolute amount. Furthermore, we generate more electricity from these basic energy sources than any other country in the world—about one-third total worldwide consumption. And in per capita terms, we also lead the world in energy consumption.

Demand for energy in the United States has also been growing steadily, at about 4.5 percent a year, while electricity generation has been surging forward at more than 7 percent a year. This means that our use of energy has been doubling about every 16 years; and use of electricity, in little more than 10 years.

Furthermore, there has been a major shift in patterns of energy use during recent years, with demand for natural gas—a cheap, clean fuel—going up, while relative demand for coal has been going down, a reflection of the rising cost of production as well as the costs of eliminating impurities like sulfur to meet air pollution standards. Together, oil (46 percent of consumption) and natural gas (32 percent) accounted for more than three-quarters of U.S. energy consumption in 1972. The dominant position of oil and gas is likely to continue under current energy use patterns and governmental policies, although the nuclear power share will rise, and the United States will probably use more coal as other fuels become limited.

These facts and projections indicate that the United States must find sources of energy by 1980 that will permit consumption to be about 55 percent higher than in 1970 (while electricity generation almost doubles), assuming that everything else remains the same. Meanwhile, worldwide consumption may increase by as much as 72 percent. These figures provide a practical basis for the so-called energy crisis.

But will everything else remain the same? What, indeed, could be done to reduce future U.S. demand for energy? How do we break the habit of mind that equates our energy *needs* with energy *demand?* These questions may be answered in several ways. Let us first consider the impact of price on the choices made by individuals and industry.

Price

Principally, the high level of energy consumption in the United States results from our high standard of living, plus low energy prices relative to our incomes and to many other resources (such as the cost of paying workers to do the same work as a machine). The price of energy also reflects conscious decisions of government, which have helped stimulate the use of energy. Thus in the United

7

States, production costs for domestic petroleum are higher than for Middle East oil imported in great quantities by Western Europe and Japan. Yet, the relatively low taxes we pay on gasoline keep the price at about 35-40 cents a gallon, compared with West European prices of between 75 cents and $1.00. Consequently, few American consumers have a financial incentive to purchase small cars that will burn less gasoline. This choice is reflected in the high level of demand for gasoline—representing most of U.S. oil consumption.

1981 Gasoline Prices

$1.40 a gallon

Similarly, the U.S. government has for many years kept the price of natural gas in interstate commerce at arbitrarily low levels in order to benefit consumers, in part because of consumer and public utility (gas) lobbying. Thus, the U.S. consumer pays far lower prices for natural gas than does a West European consumer or a U.S. industry dependent on gas that does not enter into interstate commerce.

For the same reason, the demand for regulated natural gas in U.S. industry has grown steadily over the years, and in 1970 gas was used to generate fully 28 percent of U.S. electricity. Many observers now argue that industrial and electric power generating uses of natural gas should be discouraged, in view of rapidly dwindling U.S. reserves.

In general, the use of cheap energy in the United States has helped reinforce our life-style—for example, our tendency to keep our homes warmer in winter and cooler in summer than in most other developed countries. Meanwhile, we pay relatively little attention to home insulation or the conservation of electricity that President Lyndon B. Johnson once dramatized by turning out the lights in the White House.

Consequently, many observers at home and abroad contend that we are the world's most wasteful users of energy, even though some other countries are actually less efficient in the methods they use to burn fuel. This judgment is based on some uneasiness about the large share of the world's energy resources consumed in the United States, a share which is even larger when measured in per capita terms. (U.S. per capita annual consumption in 1970 was 11.1 metric tons of coal equivalent; worldwide it was only 1.9 metric tons, a ratio

First Avenue, New York, where the gasoline goes

United Nations

of 1:5.9). But it is also based on fact. To cite three examples: no other society places so much emphasis on the use of the private automobile as compared with public transport; uses so much electricity in lighting commercial buildings through the night; or is so spread out in suburban tracts that require more energy to transport people to shops and cities. Meanwhile, the developing world, with more than 2.5 billion people, uses only about 14 percent of the world's energy—a per capita consumption of about one-thirtieth that of the United States.

Observations about the relationship between the price and

9

demand for energy in the United States do not necessarily lead to a further conclusion: that simply raising the price would effectively limit demand. Some economists do believe that a rise in the price of electricity, for example, coupled with slowing population growth, would make nonsense of projections of future U.S. demand, like those made above, and would considerably reduce worries about an energy shortage. Other economists, however, argue that raising the price of energy by only small amounts would not significantly or rapidly slow the growth of demand. Rather, they argue, the consumer ethic in American society would lead most Americans to continue buying electrical appliances and large automobiles, even if the price of electricity and gasoline were to increase. Of course, if prices went high enough, these economists agree, in time we should begin to see some slowing down of demand—though, to have an effect comparable to that in Western Europe, for example, the price of gasoline might have to go to $1.00 or $1.50 a gallon (in our larger cars) before it would command a similar proportion of our incomes. At that point, we could reasonably expect quite a significant reduction of gasoline use and greater pressure on Detroit to produce smaller cars.

Few observers have suggested that the price of energy should be raised artificially to such heights simply in order to limit demand. However, decontrolling the price of natural gas in interstate commerce may be in order if the consequent rise would limit its industrial use—where other energy sources are available—in order to conserve supplies for home use. The price of gasoline could also be allowed to go up as both domestic and foreign crude oil prices increase, thereby hopefully beginning to limit demand.

However, significant price rises would not only help to limit demand; they would also impose relatively higher costs on the poor. Everyone has a certain minimum demand for energy—for example, in home heating and lighting. And the lower one's income, the greater an impact this minimum demand has on total spending power. Thus, there is a social argument for taxing away some of the increased profits that price rises will bring to the energy industry and then redistributing added revenue, one way or another, to those people in American society who would be most affected. Of course, one reason for letting prices rise for, say, natural gas is to increase company incentives for finding new domestic sources of supply. But the Federal government also has a social responsibility to help determine what the proper level of incentives is, and to see that one group of Americans is not unduly penalized by a general rise in energy prices.

Energy Use Policy

Letting the price of energy rise in the marketplace is one way to limit demand; adopting a conscious "energy use policy" is another, as part of an overall national energy policy. The following is a list of possibilities:

1. *Rationing.* The Federal government could set up formal rationing—following the *ad hoc* rationing by some fuel oil companies during the winter of 1972-73 and by some gasoline distributors during the spring of 1973. On the grounds of equity, rationing supplies of a commodity in short supply would be desirable. However, it would also be administratively cumbersome and perhaps even unworkable (Senator John V. Tunney—D. of Calif.—has said that gasoline rationing would require the National Guard to enforce it in southern California). And a rationing system creates some inequities of its own.

2. *Taxes on gasoline.* Price rises for gasoline could come in the form of increased taxes—to 50 cents a gallon or more.

3. *Taxes on automobiles.* Sales and use taxes on automobiles could be set at steeply rising rates according to gasoline consumption, in order to discourage use of high horsepower vehicles and auto air-conditioners that increase fuel consumption between 10 and 20 percent. Or they could simply be banned.

4. *Price structure of energy.* The whole price structure for consumption of natural gas and electricity could be reversed: instead of lowering the price as more energy is consumed, the producers and distributors of energy could raise it, in order to stimulate conservation and thereby end today's subsidy of large users (like industry) by small users (like homeowners).

11

5. *Advertising.* Utility companies could be banned from promoting the use of energy through advertising, or the costs of advertising could be disallowed as a business expense for taxation.

6. *Building codes and city planning.* In building codes, greater emphasis could be placed on building design and insulation. Fully air-conditioned office buildings could be taxed. Cities and residential areas could also be planned to make more efficient use of energy.

7. *Labeling appliances.* Industry could be required to place labels on all home appliances, indicating fuel consumption, to broaden consumer choice. Frost-free refrigerators could be heavily taxed, as could gas stoves with pilot lights (both using about 50 percent more fuel).

8. *Efficiency.* Industry could be encouraged through tax or pricing policies to use energy more efficiently. Cities could use refuse as fuel.

9. *Depletion allowances.* Tax incentives for production of energy could be eliminated, thus allowing fuel prices to seek their own level. As will be discussed below, however, this possibility has other consequences that might not be acceptable.

10. *Mass transit.* Vastly greater amounts of Federal money could be spent on developing and operating mass transit systems, which are commonplace in Western Europe. However, there has long been stout resistance in Congress, and on the part of vested interests, to diverting funds from the Highway Trust Fund for this purpose. In April 1973 the House of Representatives again defeated an attempt to shift highway funds to mass transit, thus making a joke of congressional concern about an energy crisis. Most Americans also continue to resist a shift from the private automobile to expanded construction and use of mass transit within cities, or from automobiles and airplanes to surface mass transit between cities. The latter has actually declined during recent years.

Protecting the Environment

The demand for energy in the United States is also affected by concern for protecting the environment. For example, Federal standards for controlling auto emissions will tend to cut gasoline mileage, perhaps by as much as 20 percent, and thus increase consumption—provided the increased cost per mile will not lead to significant reductions in auto use.

This is not to suggest that pollution controls, either in autos or

"Paw, what's an energy crisis?"

Pearson in
Knickerbocker News

industry, should be reduced or scrapped altogether. But there is a choice that needs to be met and debated on its own terms. How much do we want to spend on protecting the environment? How important is conserving energy through relaxation of pollution standards, especially if the alternative is increased imports? And who should pay the cost? There are no simple answers. But a debate on the subject should be part of creating an overall national energy policy.

The Problem of Attitudes

In sum, the effectiveness of any energy use policy in the United States will depend upon greater awareness that limiting demand is important, either to restrain energy costs, meet problems of pollution or reduce problems of supply. Yet, as with the effect of limited price increases on the demand for energy, the American consumer ethic is dead set against attacking the energy problem from the angle of limiting demand. All too often, personal success in the United States is viewed in terms of a constantly rising standard of living; and that rise is frequently measured in terms of increased private consumption rather than public investments that could actually bring greater satisfaction and happiness. A general disillusionment with government is also increasing public unwillingness to accept a role for it in meeting social problems, even where effective efforts could improve life for all concerned.

There is also strong resistance among American political, business and technological leaders to focusing on the demand side of the problem. After all, seeking increased sources of energy is largely a matter of science, technology, price and investment. Except where there is a direct choice between two important goals—for example, between environmental protection and strip-mining, or between increased imports and national security—increasing the supply of energy involves less political and social pain than would restraining demand for energy. Thus the American proclivity to seek technological ways of avoiding messy political choices dominates thinking about a major aspect of the way we live. This approach is likely to continue, unless outside factors force a major change in attitudes—factors like steeply rising prices for *any* kind of energy; massive environmental damage; or a worldwide tailing-off of fossil fuel supplies before technological alternatives are ready for widespread commercial use.

Yet despite all that might be done, the demand side of the energy picture is still not attracting much attention or corrective action. It is necessary, therefore, to place much greater emphasis on finding new supplies of energy.

The Domestic Supply
of Energy

However one looks at it, the supply of energy is limited. The second law of thermodynamics posits the running down of the entire universe, although this phenomenon will be of little concern to policy-makers for perhaps billions of years. Solar energy, meanwhile, will be in abundant supply—unlike the earth's minerals—although the ability of mankind to make use of it is at present quite limited. Nuclear fusion—duplicating on earth the sun's releasing of energy—could sustain the earth's energy needs for centuries, although, until its secrets are unlocked, the potential fusion power contained in the hydrogen of the ocean waters will remain unused. Nuclear fission is already being used widely, but still provides but a small fraction of energy consumption.

Among all other alternatives—including water power, a relatively limited source of energy—mankind remains dependent upon energy stored up by the action of the sun during the past 350 million years. The fossil fuels created in this way are being consumed at a rapid rate—and can only be replaced at the same slow pace that has obtained throughout the ages. There is a finite limit to the earth's total supply; and, more important for this discussion, there is an even more pressing limit to the supply of fossil fuels contained within the United States.

Even so, it is doubtful that the United States would begin to face a shortage of fossil fuels for decades to come if all these fuels could be recovered for use economically. At current rates of consumption, we still have three to five centuries' worth of coal; oil shale, now hardly tapped, could provide major amounts of petroleum; and fossil fuels in unknown quantities remain to be discovered—just as Alaskan oil lay undiscovered until recently.

Yet even today there are major constraints on the availability of U.S. fossil fuel supplies. At current prices, current incentives for new exploration, and current attitudes toward pollution (in pumping, shipping, refining and burning), U.S. petroleum production—according to many experts—may already have peaked in the 48 states; and the same is likely to be true for domestic natural gas. In recent years, the "finding rate"—the ratio of wells dug to wells that strike oil and gas—has been falling; and the ratio of proved reserves to annual production fell during the period 1950-71, from 13.6 to 11.3 years' reserve capacity of liquid hydrocarbons, and from 26.8 to 12.6 years' worth of natural gas. We are now actually consuming more natural gas than we are finding new reserves.

Even with higher prices, more generous economic incentives for producing companies and less public concern for protecting the environment, most experts believe it is unlikely that U.S. petroleum and natural gas production could be expanded much further. (Some other economists believe that U.S. proved reserves could be multiplied severalfold if higher costs of recovery were accepted.) If the more pessimistic predictions are accurate, in a very short time U.S. production must slacken off (oil production fell 0.3 percent in 1972 and gas production may decline this year). Certainly, it appears that today's supplies of these two energy sources cannot long sustain a total annual growth in consumption of 4.5 percent. The discoveries in Alaska may not extend U.S. production at today's levels—about 11 million barrels/day of liquid hydrocarbons—more than several years unless, once again, today's predictions prove to be much too conservative.

The most important constraint on tapping available U.S. reserves is the cost of investment which, if passed along to the consumer, must be reflected in price. Quite simply, our most accessible reserves of natural gas and low-sulfur oil have already been tapped; each additional cubic foot of gas or barrel of oil that is added to proved reserves can be found and pumped only at increasing cost. Some relaxation of pollution standards on energy production would help, as would a faster rate of leasing offshore areas for exploration: but for how long is a matter of debate.

The Federal government already plays a critical role in stimulating domestic exploration for gas and oil, primarily through tax incentives. Benefits to energy-producing companies include the oil depletion allowance, investment tax credits, rapid amortization of investments, tax credits for intangible drilling costs and for taxes paid to foreign governments. Whether these incentives should be increased is a matter of serious, often contentious, debate. Some observers urge greater tax incentives as a critical means to increase U.S. oil and gas production, and the construction of needed refineries. Others argue that these incentives have been relatively ineffective in the past (and contend that the rate of domestic drilling actually went down following the imposition of protective oil quotas in 1959). They also argue that rising prices for energy will be incentive enough. According to this latter school of thought, any tax incentives to the energy industry should be tied closely to actual costs incurred in new explorations, not to more general incentives like the depletion allowance. Be that as it may, the issue of Federal incentives for new oil and gas exploration must be faced squarely this year.

Oil Shale and Coal

The two most important alternative sources of domestic fossil fuels are the oil shales in the West and coal. Production of oil shale could become a commercial proposition in large quantities (perhaps 600 billion barrels, or more than today's total world proved oil reserves). But this would require large amounts of investment capital—some $3 billion for the first 1 million barrels/day of pro-

Oil Shale development program in Parachute Creek, Colorado.

duction or one-seventeenth of current consumption; higher prices for oil; sufficient water; and less emphasis on environmental problems.

The use of coal is also a matter of investment and price, especially to develop technologies either to (1) liquify and desulfurize coal for use as petroleum; (2) desulfurize coal for electricity generation; (3) gasify coal into methane, the basis of natural gas; or (4) use coal to produce hydrogen, which then can be burned cleanly. The Federal government is already investing in some coal technologies in support of industry—but still only at the rate of several miles of superhighway each year. In addition, techniques are being developed to reduce the sulfur and other polluting products in stack gases, but, again, at a price.

This last example points to a major cost of using our reserves of coal. Pollution standards in force today and projected for tomorrow impose considerable costs that can be measured in dollars. The social and physical costs to the country of burning high-sulfur fuels would also be high, even if they cannot be quantified with the same ease. Similarly, economical methods of getting at coal today center on strip-mining, with its spoilage of lands used for this purpose. Replacing the soil is theoretically possible—once again at a stiff price and with residual damage.

New Sources of Energy

Besides concentrating on finding new reserves of fossil fuels and methods of exploiting proved reserves, the United States has a number of other ways of meeting some of its long-term energy needs from domestic sources. Indeed, it is often argued that our domestic energy crisis will be only of a few decades' duration in any event, *provided* that new technologies open up the possibilities that are now foreseen. The two most important possibilities involve nuclear power.

1. *Fission.* The use of the nuclear reactor for production of electricity has been known for some time. Although electricity generated from nuclear reactors in the United States now provides only about 0.6 percent of our total energy consumption, this process will expand rapidly in coming years.

Today's nuclear reactors are of limited long-term value, however, because they consume all the nuclear fuel that is put into them. However, development is now under way on a much more efficient use, both here and abroad. Indeed, the United States may lag behind the Soviet Union and some West European countries in this technology, leading to some criticism that the U.S. has neglected research on several alternative fission reactors, including the gas-cooled breeder. Instead most research and development effort has been concentrated upon the Liquid Metal Fast Breeder Reactor (LMFBR), which relies upon a quirk of nuclear fission to produce more fuel than it consumes.

At present, the Federal government is investing substantially in development of the LMFBR. A prototype should be ready by 1980, though the LMFBR will only become an important source of electricity later in the decade. By the beginning of the next century, the fast breeder reactor may be producing a major fraction of U.S. energy needs. This will be none too soon: for some uses (such as lubrication, manufacture of chemicals and the automobile as we now know it), fossil fuels cannot easily be replaced. Of course, we should still have ample coal reserves for these purposes, and we might by then have auto engines running on hydrogen fuel cells.

The nuclear reactor has two serious technical drawbacks, however: first, it produces large quantities of waste heat, which must somehow be dissipated. With current and projected technologies, this requires water—and lots of it. Already, the environmental impact of reactors has been felt in many parts of the country; and there is increasing public uncertainty about the safety of these plants. (One report about a reactor incident noted that "100 percent of subject biota exhibited mortality response," by which it meant to convey that all the fish died.) Thus, there are real limits on the use of nuclear fission to generate electricity. The same reservation applies to the second drawback in fission reactors: the production of radioactive by-products that retain their potency for hundreds or thousands of years. These can be disposed of in many ways, but any way leaves behind a potential hazard that will be with us for centuries to come.

2. *Fusion.* The answer to the scientists' prayer is the prospect that nuclear fusion—the process that powers the sun—can be used

commercially on earth to produce energy and generate electricity. If it does become practicable sometime early in the 21st century, energy supplies will theoretically be available until the far distant future. Water itself contains a concentration of between .014 and .015 percent of the hydrogen isotope deuterium, one part of the fusion reaction.

The technology, however, is incredibly difficult. Whereas the fission reaction can be controlled with relative ease, true fusion has only been experienced on earth in the fireball of the hydrogen bomb. To produce a controlled source of energy, fusion requires temperatures of between 100 million and 1 billion degrees, maintained for hundredths of a second (aeons in nuclear time).

Other technologies are being explored. The two most favored are the use of high temperature gases (plasmas) circulating within magnetic fields; and the bombardment of "fuel pellets" with high-powered laser beams. No controlled fusion reaction has yet been sustained; it may be several years and require enormous investments before a laboratory design is perfected. Even when fusion power becomes available, it will be subject to many of the constraints, other than availability of fuel, besetting fission power and the burning of fossil fuels. Although fusion will be both relatively clean in burning and waste disposal, it will still raise problems of waste heat dissipation.

Alternative Energy Sources

As always, scientists are proving inventive and resourceful in the search for new energy supplies. Several possibilities are listed here, and they by no means exhaust the list:

1. *Solar energy.* In the not-too-distant future, direct use of the sun's radiant energy could become practicable on a large scale for space heating, although solar energy for generating electricity is much farther off. Solar energy is already used in home heating in some places and has proved economical. At some point, we may also see the building of large "solar farms," either using radiant energy directly for electricity steam generation or production of hydrogen,

21

or storing the energy in molten salts or as hydrogen for nighttime use. The necessary size of these farms, however, is subject to debate and, hence, so are the capital costs involved. Even so, large desert areas could well be suitable.

Solar energy has the technical advantage of being clean compared to fossil fuels. And it has two political advantages: (1) it is available to many countries without import or export; and (2) unlike nuclear reactors, it uses no fuel and produces no radioactive by-products that could be used for weapons manufacture. By 1980, for example, a large number of countries will be in a position to build an atom bomb from materials in nuclear reactors designed to produce electricity. Of course, the nonproliferation treaty provides for some safeguards against the diversion of nuclear materials for weapons use; but countries decide to make atomic bombs essentially for political reasons. Yet the widespread use of nuclear reactors to generate electricity does increase the technical opportunities open to many nations to make this choice and increases the desirability of alternative energy sources like solar power.

2. *Geothermal energy.* It is also technically possible to make use of the heat trapped beneath the earth's surface. In California, 300 megawatts of electricity are already being produced from a natural geothermal steam field; alternatively, water can be pumped down a deep hole, to reemerge in a companion hole as steam for electricity generation. With current technology geothermal sources could meet about 1.5 percent of total U.S. energy needs by the year 2000. Critics warn that the progressive, though limited, cooling down of the part of the earth just below its crust would threaten earthquakes

3. *Other methods.* Other ways of producing energy include the following: (a) magnetohydrodynamics, or the use of hot hydrogen gas to generate electricity from magnets; (b) tidal energy (a very limited source); (c) giant windmills; (d) underground nuclear explosions to heat water into steam; (e) space collection of solar energy, to be beamed to earth from satellites; (f) burning refuse; and (g) growing trees on refuse and burning them as charcoal.

It should be stressed that the range of American choice will not be determined by price (of one kind or another) alone. Even more important at the moment is the matter of lead time: whether new technologies, or new explorations, could eliminate the short-fall in U.S. energy production during the next several years, whatever industry and government choose to do today. The answer seems to be a clear No. Indeed, some increase of short-term U.S. dependence on foreign sources of energy supplies appears inevitable, even if prices continue to rise, investments are increased massively and environmental standards are completely scrapped. Senator Henry M. Jackson has proposed a crash program of new investments in energy, with $20 billion or more of Federal money over ten years. This proposal, along with others for a more diversified development effort, should be given serious consideration as part of an overall energy policy for the future. But for now, it is simply too late to take care of today's growing gap between U.S. domestic energy supply and demand. What this means will be explored below.

Importing Fossil Fuels

For the United States, the trends of domestic demand and supply of energy discussed above indicate that we will have to import far more energy from abroad during the next few years than ever before, regardless of what steps we take now. This year our imports of oil may rise to 6 million barrels/day, over one-third of consumption (compared to 26 percent two years ago). Our dependence will remain considerable, even if we elect to bear the high financial and environmental costs of promoting as much self-sufficiency in fossil fuels as possible—costs that should be the subject of public debate as we consider quotas and tariffs designed to limit imports.

In 1980, with unrestricted imports and no change in other U.S. policies or the rate of growth of U.S. demand, we might import 10 million to 12 million barrels or more of oil a day, out of total consumption of over 22 million to 26 million barrels/day. According to the best estimates available, only in the long term—a matter of one to three decades, which is beyond the "time horizon" of any politician or planner—could our foreign dependence be reduced without draconian limits on domestic demand.

This conclusion is agreed to by virtually every observer of the U.S. energy picture; indeed, the figure of 50 percent or more of U.S. oil consumption from abroad by 1980 has almost come to be dogma. For example, this view was supported in 1971 by the National Petroleum Council (NPC). Its study forecast dependence on foreign oil in 1985 running as high as 57 percent of our consumption; and on foreign natural gas, as high as 28 percent. By the time the NPC's summary report was issued in December 1972, its estimates had increased and projected a 65 percent dependency on oil. More conservatively, the Interior Department has speculated that we will depend on outside sources for 25 percent of our gas and 40 percent of our oil during the 1980's.

Of course, estimates like these assume a continuing increase in U.S. demand, a sluggish response of new domestic exploration to price rises or new investments and little increase in the availability of synthetic fossil fuels. But even with more optimistic assumptions, the extent of our foreign dependence will be considerable.

To be sure, the United States has for some time been a net importer of energy, except coal, and today it gets more than 10 percent of its total energy from abroad. This compares with a net export of energy in 1925 of about 3 percent and a net import in 1951, the year of turnaround, of 1.5 percent. In oil, the shift has been even more dramatic: from 8 percent net import in 1950 to a net import of about 29 percent in 1972. (Our total energy deficit is rising less fast than that of oil, because of increased exports of coal.)

Table 2

Oil Imports

(million barrels/day)

	1972	Est. 1973	Est. 1980
United States.............	4.7	6.0	10–12
Western Europe..........	14.4	15.5	22–26
Japan...................	5.0	5.5	10–13
TOTAL.................	24.1	27.0	42–51

Source: *Presidential Message Fact Sheet*, April 18, 1973.

Table 3
Sources of Crude Oil and Oil Product
Imports into the United States

Country or Region	thousand barrels/day			percent		
	1965	1970	1971	1965	1971	1973[1]
Canada	325	690	800	13.2	20.6	20
Caribbean	1,565	2,135	2,170	63.5	56.0	
Other Western						38
Hemisphere	60	5	30	2.4	0.8	
Western Europe	5	210	145	0.2	3.7	
Middle East	350	175	390	14.2	10.1	
North Africa	55	75	90	2.2	2.3	
West Africa	15	50	105	0.6	2.7	
Southeast Asia	90	70	130	3.7	3.4	42
U.S.S.R., Eastern						
Europe	—	5	5	—	0.1	
Other Eastern						
Hemisphere	—	5	10	—	0.3	
World	2,465	3,420	3,875	100.0	100.0	100

Source: British Petroleum Company, *Statistical Review of the World Oil Industry*
(published annually).
[1]*Presidential Energy Message Fact Sheet*, April 1973. Current Rate.

Until now, however, the U.S. import of oil has been concentrated largely in the Western Hemisphere: from Canada, about 20 percent in 1971, and from the Caribbean, 57 percent in 1971, which largely means Venezuelan oil, as well as some Eastern Hemisphere oil that is refined elsewhere in the area. In the future, however, the size of U.S. oil and natural gas imports should change dramatically.

Sources of Imports

What degree of choice does the United States have in selecting its sources of foreign energy? This is a critical question for foreign policy. Its answer requires a look at the world's reserve picture.

At the moment, worldwide proved oil reserves—the only usable measure of untapped reserves—continue to increase. Today, the non-Communist world has proved reserves of 550 billion to 600 billion barrels, or the equivalent of 36.7 years at 1972 rates of world-wide production. This is actually an increase from 32 years' worth of proved reserves in 1965-67. Again, however, these figures must be viewed with caution. If anything, they underestimate the world's proved reserves; some governments and companies are reluctant, for reasons of bargaining, to produce true figures; and some discount reserves that cannot be recovered profitably at prevailing costs and prices of energy. Nor do we know as much as we need to know about future supplies of energy, or even where as-yet-undiscovered oil and natural gas are located.

At present, proved reserves are distributed unequally around the earth's surface, with at least three-fifths of them located in the

Table 4
Worldwide 'Published Proved' Oil Reserves, end of 1971

Country/area	Barrels (billion)	Share of total (percent)
United States[1,2]........................	45.4	6.8
Canada[2].............................	10.2	1.5
Caribbean............................	17.1	2.8
Other Western Hemisphere.............	14.5	2.3
Total Western Hemisphere...........	87.2	13.4
Western Europe......................	14.8	2.3
Africa...............................	58.9	8.9
Middle East..........................	366.8	57.6
U.S.S.R., Eastern Europe, China........	98.5	15.4
Other Eastern Hemisphere.............	15.6	2.4
Total Eastern Hemisphere............	554.6	86.6
World (excluding U.S.S.R., Eastern Europe, China.....................	543.3	84.6
World...............................	641.8	100.0

[1]Including estimated North Slope reserves of 9.6 billion barrels.

[2]Includes oil that can be recovered for natural gas.

Proved reserves are generally taken to be the volume of oil remaining in the ground which geological and engineering information indicate with reasonable certainty to be recoverable in the future from known reservoirs under existing economic and operating conditions.

The data exclude the oil content of shales and tar sands.

Source: British Petroleum Company, *Statistical Review of the World Oil Industry 1971* (1972).

Middle East and North Africa—almost one-fourth in Saudi Arabia alone (its actual reserves may be several times greater). Reserves there and elsewhere will no doubt expand, but by how much is still quite unclear. A few years ago, for example, the North Sea was unheard of as a source of supply (as was the Arctic); in the near future, it is expected that this field will permit some European countries to reduce their dependence on foreign sources of petroleum, though for how long is subject to debate.

For purposes of this discussion, it is necessary to work with figures of proved oil reserves that are now publicly available, recognizing that even if some reserve figures prove to be grossly underestimated, the lead time to develop sources of supply is still a critical factor.

To begin with, there is certainly a strong emotional incentive on the part of many Americans to seek foreign energy supplies from within the territory covered by the Monroe Doctrine. At the moment, this means essentially Venezuela and Canada, because of the size of their reserves.

Venezuela

Venezuela, which has a limited domestic need for oil (in contrast to Canada), is the more likely future source of U.S. petroleum imports. For all practical purposes, however, Venezuela's production has peaked out. Venezuela is not necessarily running out of oil. The government in Caracas does not really know how extensive its reserves are, especially since foreign company explorations were brought to an end during the 1960's.

Rather, Venezuela's attitudes toward the role of energy in its economy are changing. In the first place, present agreements between the government and the companies (largely Royal Dutch-Shell and Creole, a division of Exxon) expire in 1983 and are unlikely to be renewed, even on terms more favorable to the government. This prospect has reduced the incentives that foreign companies have to invest further in Venezuelan oil. More important, there are strong pressures in Venezuela toward conservation of a resource that must by definition be finite. Preserving a stable source of revenue for the future, rather than overproducing for the short term, is clearly an important factor in Caracas' approach to oil policy.

There is also strong opinion among some political interests in Venezuela that oil has distorted the nation's economy, creating a concrete jungle of superhighways in Caracas that may soon rival those of Los Angeles, while this tropical nation actually imports some fruits and vegetables. Consequently, there is a search for a more balanced economy, one that will have some chance of solving critical problems of unemployment, rural-urban migration and erosion of the agricultural sector. In the process, restraint on the growth of oil revenues may actually be desirable.

For all these reasons, the United States cannot count on Venezuela as in the past for a major portion of its oil imports, even if prices rise (under some circumstances price rises actually impel the move toward restricting output, since the same revenues can be had for less oil pumped). And the reasons for this restriction are compelling ones; not because of a lack of potential reserves, which some observers believe to be as much as a *trillion barrels* of heavy oil, but rather because of the relation between oil economics and the overall development of the Venezuelan economy. (Of course, the costs of recovering the heavy oil—technically difficult—could be immense, reaching $5 billion to $7 billion in investment for every 1 million barrels/day capacity.)

Canada

Canada is the other most desirable source of petroleum for the United States. Its oil industry is relatively young and undeveloped. Yet its potential is considerable. Although Canadian proved oil reserves at the end of 1971 were only about 10.2 billion barrels (less than the equivalent of only 2 years of current U.S. consumption), some experts believe they could expand greatly. In addition, the Athabasca tar sands of Alberta and Saskatchewan could alone yield perhaps 300 billion barrels of oil, 50 billion barrels of it with relative ease, but the rest at enormous investment cost. Also an unknown quantity of oil and gas is located in the Canadian Arctic.

At a first glance, it would appear that large supplies of Canadian energy would be available for the U.S. market. Agreements already exist for the supply of natural gas and some oil; eastern Canada—which receives almost no Canadian fossil fuels from the West for economic reasons—actually gets some Middle Eastern and Caribbean oil through the United States; agreements exist for the supply of hydroelectric power to the eastern United States; and American firms play a dominant role in the Canadian fossil fuels industry. These developments have led many Americans to talk about a "continental energy policy," embracing the United States and Canada in a sharing of resources.

The reaction in Canada is quite different. For example, there is widespread concern about the environmental consequences of bringing fossil fuels from the North (including Canada's own North Slope)—especially to serve a U.S. rather than a Canadian market. There are concerns about conservation and about future self-sufficiency in fossil fuels. The Canadian Energy Board has even forecast a decline in oil production in 1977 and an oil shortage by 1986.

Most important, however, the Canadian government, with the support of most of its people, is resisting any notion of a continental energy policy in which Canada would once again be treated as a junior partner of the American colossus. Too much of Canadian industry is already owned by Americans for comfort north of the border; and economic relations between our two countries have been far from harmonious in recent years. It is most likely, therefore, that Ottawa will see the American energy crisis as an opportunity to reverse in part the economic partnership that has worked to the disadvantage of Canada's interests in maintaining control over its domestic economy. Many Canadians, therefore, look forward to U.S. energy problems with equanimity—if not some glee. Any bargain that is struck will come very dear indeed and would probably focus on gaining more favorable terms of entry to the U.S. market for Canadian manufactured goods.

Yet for Canada to use energy as a bargaining lever with the

United States would require a reversal in current Canadian policy. Since November 1971, for example, Canadian natural gas exports to the United States have been held to the levels then obtaining. On February 15, 1973 the Department of Natural Resources imposed controls on the export of oil to the United States and implemented these in March by limiting the shipment of oil to 1,235,000 barrels/day, 47,598 barrels/day less than pipeline companies wanted. At the time, the Canadian government listed a lack of pipeline capacity as the reason for what is seen to be a temporary restriction, with Canadian needs being put first.

For the United States to expect increased energy supplies from Canada would require at the very least a major change in U.S. attitudes toward our closest neighbor. New attitudes will not be easy to create in view of a long history of relations in which the U.S. has tended to patronize Canada. New attitudes are essential, nonetheless, as part of long-range economic cooperation. They are essential, even though the amount of Canadian fossil fuels that might in any event be shipped to the United States would only be a small fraction of U.S. consumption, except in the unlikely event that major exploitation of the tar sands were pushed forward with vigor. Thus Canada, too, provides no real or lasting answer to the U.S. energy problem.

Sources Outside the Middle East and North Africa

The United States has a number of other potential sources of oil imports besides Venezuela and Canada outside the Middle East and North Africa: In the Western Hemisphere, Colombia expects to find and export significant quantities of oil, and some discoveries have been made in other South American countries like Ecuador and Peru. However, to obtain large quantities of oil (and natural gas), the United States must definitely go outside the Western Hemisphere.

Two other countries stand out: Nigeria and Indonesia. Nigeria currently has about 2 percent of worldwide proved reserves, but expects to find much more, and has a substantial and growing need for income to finance its economic development. In 1972 it pumped more than 1.8 million barrels/day, a one-year increase of 17.2 percent (reflecting, in part, the end of the war with Biafra). At some point, therefore, Nigeria could produce enough oil to become a significant factor in world energy trade. However, much of this oil will be destined for the European market despite growing U.S. imports, and Nigeria is most unlikely ever to rival the major Middle Eastern and North African producers.

Indonesia also has a tremendous need for development capital. So far, however, its proved reserves amount to less than 1.5 percent of the worldwide total, although exploration still has a long way to go and production is already over a million barrels/day. It may rise to 2 million to 3 million barrels/day by 1980. Here, the United States would face strong competition from Japan. It now takes the bulk of Indonesian oil, although in 1972 the United States did import about 150,000 barrels/day of Indonesian oil. The United States might have continued access to some of this oil, especially if the price of alternative energy supplies were higher in the United States than in Japan. In addition, Indonesia has a general concern to diversify its economic ties with other nations and, indeed, several U.S. oil companies are already operating there.

However, the production and export of Indonesian oil will also be seen in Djakarta within the overall political context of that country's relations with the outside world. Gaining access to Indonesian oil in significant quantities—and even then only a small fraction of U.S. imports—may also require the United States to pay much greater attention to Indonesia's problems of development. But if the United States acts intelligently within this context, it stands to gain some benefits from Djakarta's energy policies as well.

The Middle East and North Africa

All the previous alternatives, widely if not always accurately seen in the United States to be politically "safe," still could not provide enough oil under current policies and with current projections of domestic supply to meet projected rising U.S. demand. Like it or not, in the next several years, at least, the United States will have to rely on the Middle East and North Africa for considerable quantities of oil, as well as of natural gas, particularly now that it is technically and economically possible to liquify natural gas at low temperatures or convert it to methane and transport it by sea. This also means that in the short-term the United States and other consumer countries must cope with all the political and economic problems of depending on the countries of this region for more oil.

The United States is only now becoming heavily involved in importing fossil fuels from the area. During the latter part of the 1960's, the Middle East provided between 5.1 and 14.2 percent annually of total U.S. imports of crude oil and products. In 1971, the Middle East and North Africa accounted for 12.4 percent of total U.S. oil imports (3.2 percent of U.S. oil consumption). However, this share of imports is beginning to rise steeply, and may reach 50-55 percent or more by 1980 if there are no limits imposed on U.S. imports or a halt to the growth of U.S. demand. Meanwhile, most of the producer states are increasing their production. Iran expects to reach 8 million barrels/day and then level off (although it might go to 10 million barrels/day until production begins to tail off in the 1980's). Saudi Arabia plans to produce 20 million barrels/day—a staggering (and probably unrealistic) increase from today's 6 million barrels/day. This figure should be compared to today's total U.S. daily consumption of about 17 million barrels. Kuwait has exceeded 3 million barrels/day although it has now imposed a 3 million barrels/day limit for purposes of conservation. Iraq may reach 5 million in a pinch; the other Persian Gulf states may reach 6 million; Libya, 2 million (a reduction from present-day levels for conservation and political reasons); and Algeria, 1.5 million. Meanwhile, Algeria will also be producing large quantities of natural gas, and by 1980 may supply as much as 5-10 percent of U.S. domestic consumption and a much larger percentage of East Coast consumption.

If these projections have any validity, they add up to 45.5 million barrels/day in 1980—or more than four times current U.S. production and about twice today's production in the area. Of this 45.5

Oil rig operating in North Sea

million barrels/day, the United States might be importing 6
million to 8 million barrels or more, out of total imports of perhaps
10 million to 12 million and consumption of 22 million to 26 million.

Issues for Foreign Policy

The magnitude of these figures, along with growing U.S. dependence on North Africa and the Middle East for energy imports, raises a number of important foreign policy issues. These include: the role of the Soviet Union in the Middle East and the evolution of U.S.-Soviet relations as this bears upon energy problems; the impact of the Arab-Israeli conflict and U.S.-producer state relations generally on the supply of oil; the increasing role oil producing states will inevitably play in the international trade and monetary system; cooperation among consumer states; paying for energy imports; and, finally, the environmental dimension of increasing trade in fossil fuels.

In any consideration of what are essentially security problems of one kind or another (including financial and environmental security), it is important to bear in mind that all threats must be seen in relative terms. In the past, security has been interpreted all too uncritically. The 1959 Mandatory Oil Import Program, for example, was justified in terms of security, when its object was really to protect the U.S. oil industry against foreign competition. Nor do scare tactics have any place in today's debate about increasing U.S. energy imports from abroad. Throughout any consideration of potential threats, we must ask several questions: How serious are these threats? What can be done about them? And how much is it worth to us to seek alternatives that would bring about less energy dependence on the outside world? These are the issues we shall consider in the following chapters.

The Role of the Soviet Union

Until recently, any survey of U.S. dependence on the Middle East for oil would have concentrated heavily on the political problems of Soviet involvement in the area. Yet recently this element of the problem has been less evident in U.S. commentary, even on the part of many observers who believe that the foreign political problems of importing energy from the Middle East are critical ones. Why this is so will be discussed below. First, however, it is necessary to sketch the Soviet position in energy.

As we might expect, Soviet energy statistics are little known, at least publicly, in the United States. Indeed, the British Petroleum Company's *Statistical Review of the World Oil Industry,* issued yearly, lumps Soviet proved reserves together with those for Eastern Europe and the People's Republic of China. These are estimated at 98.5 billion barrels of oil in 1971, of which perhaps 75 billion are in the U.S.S.R. Yet some tentative conclusions can still be deduced from what the Soviet Union actually does.

To begin with, the Soviet Union still seems to have a much wider range of choice than does the United States in deciding whether or not to import energy from abroad. This reflects a considerable pool of fossil fuel reserves and vast unexplored areas (though many are beneath the Arctic tundra). It also reflects an economy still far less dependent than Western economies on forms of energy—i.e. oil—that can be used in automobiles. Nor is it evident that the Soviet Union yet stresses the environmental effects of high-sulfur fuels to the extent that the West does.

Whatever the facts of internal energy development, the Soviet Union has modified its long-standing policy of remaining aloof from trade in energy and has begun importing oil and gas from the Middle East. One reason for doing so is simply economic: transportation costs within the U.S.S.R. are thereby reduced. In addition, imports increase the amount of Soviet domestic supplies to be released for export.

Some of this energy export has at times been clearly political. Partly through this the Soviet Union has sought to extend its domination of the East European states. It has helped the U.S.S.R. to gain considerable economic leverage, as well as pay for finished products from the East European states. It can be assumed that this practice will continue whenever possible, although at the moment the Soviet Union appears to favor the economic benefits of gaining hard currency through energy sales to the West. In response to this Soviet effort, every East European state some time ago sought to enter into direct agreement with Middle East suppliers, at least in part as a way of minimizing dependence on the Soviet Union.

The Soviet Union has also made significant inroads into Western European markets, both in oil and natural gas—fuels which it ships in quantity, especially to Austria, Italy, West Germany and France. This policy is, to some extent, political—to create some dependence; but it also has important economic aspects in view of the Soviet economy's great hunger for high-technology goods from the West. The Soviet Union has the advantage of being able to pipe natural gas to parts of Western Europe—such as Italy—which might otherwise have to rely heavily on much more costly liquified natural gas (LNG). It is also willing to sell energy in the West at favorable rates, by accepting an exchange rate for the ruple far below its official value—apparently valuing the pipelines and hard currency it gets in return much more highly than sustaining the ruble at its inflated level.

Furthermore, the extent of growing Soviet involvement in Western energy markets is evidenced by a few *ad hoc* swap arrangements in petroleum to meet delivery shortages, for example, with British companies. The Soviet Union is also marketing oil directly from petrol stations in Britain. And there are Soviet-owned refineries in Belgium and France.

Beyond Europe, Moscow and Tokyo have for some time been discussing Japan's construction of a pipeline from Siberia to the Sea of Japan, to be repaid in natural gas. And there is the prospect of a major U.S. purchase of as much as several billion dollars' worth of Soviet LNG, about equal to the amount that is involved in current U.S. agreements with Algeria. This last-named development reflects the political objectives of détente on the part of both superpowers more than it does economics, since the landed price of Soviet LNG will be about three times the current sale price of U.S. natural gas in New England.

All these developments have fairly straightforward strategic implications. On the one hand, it remains important for all Western nations, and especially for the United States, to keep dependence on the Soviet Union for energy, as for any other strategic commodity, as low as is necessary to be consistent with security—i.e. low enough so that an abrupt and total cutoff of energy supplies from the Soviet Union would not have a serious effect upon availability of energy. Soviet supply of energy to Western Europe is already raising this issue; in time, U.S. imports of LNG could raise it as well, although projected imports represent only about 1 percent of total U.S. energy consumption.

On the other hand, some Western (and also U.S.) dependence on the Soviet Union for energy supplies is tolerable, and might even be advisable as a way of diversifying our sources of energy, provided there were countervailing dependence in return—dependence reflected in the Soviet need for hard currency and technology. Over time, as well, the changing nature of U.S.-Soviet relations may make possible a degree of mutual interdependence that we have so far been unable or unwilling to contemplate.

Some observers have even suggested that the Soviet Union might become a more reliable long-term source of energy than some of today's producer states in the Middle East. This idea is difficult to credit, however, on several grounds. In simplest terms, it is unlikely that the Soviet Union would be able to supply the United States with large amounts of energy. And even if a common Soviet-American desire to preserve and strengthen other aspects of a relationship that is more important bolstered U.S. confidence in the Russians as energy suppliers, we would still need to preserve and strengthen relations with other producer states, including many in the Middle East and North Africa. The Soviet Union thus offers no easy way out of a potentially difficult political problem.

The Soviet Union in the Middle East

Most important is the political role that Moscow might choose to play with regard to Middle Eastern oil and natural gas (and, perhaps, Indonesian oil, as well). It has been an axiom of Western assessments of Soviet behavior that the Kremlin subordinates economic to political objectives. With respect to energy—especially in the Middle East—it is often argued that the Soviet Union would welcome any leverage it could gain over Western supplies, and thus over Western economic security: the so-called hand on the tap. Indeed, despite Moscow's military withdrawal from Egypt during 1972, it retains considerable position and influence in a number of Middle East countries, especially in Iraq and relatively oil-poor Syria. It has also expanded its diplomatic activities in the area, generally, and its involvement in Middle Eastern energy, specifically. It has long had energy agreements of one kind or another with a number of countries, notably Iraq and Iran, that have increased its stature in the area.

On the surface, therefore, it seems that the Soviet Union is in a position to act against Western interests in energy. And it is possible that Moscow might at some point want to be able to shut off energy supplies to the West, and particularly to the United States. However, the methods by which the Russians might try to achieve this objective pose various problems. Direct invasion is out, to begin with, unless the Russians were able and willing to secure domination of the entire region at one go. Use of military power against only one or two oil producers would simply destroy the possibility of productive relations with all other states in the region and would quite likely drive them, against their inclinations, toward closer ties with the United States. Along with nations in the West, the Soviet Union must also accept the modern limits on "gun-boat" diplomacy outside its own province in Eastern Europe.

Fostering sabotage of oil investments or sponsoring regimes hostile to the West is another and more plausible possibility. Here, too, however, the hand of the Soviet Union would have to be concealed for the effect to be, on balance, one favorable to Moscow with any oil producer. None of them would accept having its option to sell oil foreclosed.

Furthermore, oil has so far generally been thicker than ideology in the Middle East. Even Iraq, with a succession of regimes since 1958 that have been more or less hostile toward the West, has ceased production for only short periods of time (though its production has often been reduced). The nationalization of the Iraq Petroleum Company (IPC) in 1972, plus long-standing energy relations with the Soviet Union, has not led to an ideological boycott of oil exports to the West; and Iraq has already reached agreement with IPC on compensation for the nationalized concessions.

A more subtle policy would be for the Soviet Union to buy up energy that might otherwise go to the West. The question of becoming too dependent on the Soviet Union for a marketing function—where Middle Eastern energy either would be shipped across the Soviet Union to Western Europe or would release domestic supplies for that purpose—has been raised above. So far, the oil-producing states have not objected to this relatively minor Soviet role. But if the Soviet Union tried to expand it, to the point of having serious leverage on supplies to the West, there would likely be resistance in the Middle East. Whatever may be predicted, most producer states are still interested, among other things, in maximizing revenues. They want to expand markets, not simply allow the Soviet Union to take them over by acting as an unnecessary marketing agent. This would become increasingly evident as producer states gained a greater equity share in (and control over) production and, in some cases, extended their role into "downstream" operations of transport, refining and marketing.

Furthermore, in view of the production plans of several Middle Eastern producers, the Russians would also have to be willing to incur a severe and sustained economic loss to use the purchase of energy as a way of gaining a hand on the tap and then using it as a political weapon against the West. This is hardly plausible. Finally, if for whatever reason major producer states did cut back shipments to the West, making Soviet energy exports more significant, the crisis would be so serious that curtailed Soviet oil exports would not be very significant. In fact, during the Arab oil embargo against Britain and the United States in 1967, the Russians actually *increased* their export of oil to the West.

In any event the development of a Soviet role in Middle Eastern energy has to be seen against the background of emerging U.S.-Soviet relations. This is a time of mutual deterrence and of détente. It is also a time when both the United States and the Soviet Union have learned to behave with one another in a much more mature and cautious way and to take few risks anywhere of provoking a nuclear confrontation.

Given this new relationship, the superpowers have less to worry about one another even in an area, such as the Middle East, where the extent of each other's influence has not yet been settled.

As far as the supply of energy is concerned, the developing context of superpower relations suggests that neither side would run even a small risk of jeopardizing much larger and more important relations in order to gain such a relatively minor form of leverage. The Russians might like to have that *potential* (and we might, as well); and the United States must not depend upon Soviet good intentions in the field of energy any more than it does in other areas. But the chances that the Soviet Union would actually try to use its involvement in Middle Eastern energy as a way of causing the United States real harm have certainly been reduced by broader diplomatic and strategic developments. Indeed, the Russians will need to be cautious of even raising *fears* in the West that they might act irresponsibly in Middle Eastern energy if they do not want energy problems to affect more serious matters of détente.

Western understanding of these matters has been increased by what so far has been fairly conservative Soviet behavior regarding Middle Eastern energy and in their trading relations throughout the world. It is likely that this conservatism will increase as the Russians become even more subject to the vagaries of Middle Eastern politics (as in Egypt) that Western countries have long faced. In time, the Russians may share Western interests in stable supplies and prices, and may face similar problems as potential neoimperialists, regardless of ideological affinities with particular producer states. They also need now to assure producer states of their conservative behavior, whatever their long-term political intentions. If there is to be any instability in marketing, producer states want this to be their choice, not that of a customer.

Of course, in time the Soviet Union may increase its involvement in Middle Eastern energy simply because it needs to do so. Predictions vary greatly. But it does appear that the Soviet Union will—perhaps by 1980—need to import more energy for its domestic use for economic reasons.

Here, there is a real possibility of future competition between the Soviet Union and the Western states for Middle Eastern energy supplies. How large or intense this competition might be is a matter for speculation. But the possibility that it could come about must be an added calculation in predicting the future price of energy and in estimating the bargaining power of Middle Eastern producers. Indeed, the Soviet Union will almost certainly become an extra party to be played off against the Western countries and companies in setting the price and terms of energy delivery. It is often argued that the Russians could not stand the competition, especially because of the inferior quality of their goods and equipment, and because of their need at some point to begin paying hard currency chronically in short supply (as appears to be happening now in Iraq). But if the Russians elect on economic grounds to import more Middle Eastern oil and gas, then they could become effective competitors.

An Exxon photo

Refinery at Ras Tanura, Saudi Arabia, biggest oil port in the world

Problems of Transport Security

Finally, in considering the role of the Soviet Union in energy, we should note the growing concern in some quarters over the safety of tankers carrying oil from the Persian Gulf for East and West coast ports in the United States. Implicit in this concern is the memory of World War II, when loss of rubber supplies in Southeast Asia posed a grave danger until the practical development of synthetic rubber, ironically, from petroleum.

Of course, we have nothing to fear for energy imports from

Canada and little cause for concern over imports from elsewhere within the Western Hemisphere. Indeed, the greater part of U.S. imports today comes from within the hemisphere, in part for reasons of security. But can we rest easy when there are vast tanker fleets carrying oil and natural gas to the United States from the Eastern Hemisphere, and especially from the Middle East? In times past, such a situation would have been a major cause of alarm during any crisis in which we were involved. Today, however, there is less to fear.

The only real threat at sea to U.S. tankers for the foreseeable future could come from the Soviet Union. On the face of it, this threat appears to be significant, especially in view of the steady growth of the Soviet navy, and particularly its submarine capabilities.

In today's world, however, and surely in tomorrow's, threats to vital supplies of U.S. energy could not be divorced from a larger confrontation. Indeed, the more vital shipments become to us, the more likely it will be that threats to them would cause a deterioration of relations with the Soviet Union and represent a major step in escalation. Furthermore, as argued earlier, both the United States and the Soviet Union have extended the areas of the world and types of action where they will take few risks.

Of course, some observers argue that the United States will need to maintain naval strength, particularly in the Atlantic, if only to preserve the psychological requirements of deterrence and to warn the Russians of the risks in threatening U.S. energy supplies. In general, however, the protection of commercial shipping and freedom of the seas will have to depend upon political factors and relations rather than upon sea power. Japan already appears to have reached this conclusion, even though 83 percent of its oil comes from the Middle East and passes through straits under the control of other countries. For the United States, therefore, the lesson is clear: as with other political approaches to a Soviet role in the U.S. energy picture, we must increasingly focus on the political context within which we will be shipping supplies across the seas.

The Arab-Israeli Conflict

For many years the absorbing passion of all countries in the Middle East—other than Iran and Turkey—has been the Arab-Israeli conflict. It has resulted in three wars, innumerable skirmishes, enormous military expenditures and inflated rhetoric on both sides. Outside powers, primarily the United States and the Soviet Union, have played critical roles in its ups and downs.

Until recently, oil figured very little in our view of the Arab-Israeli conflict. It is true that American political rhetoric often rang with charges that "Israel will be sold out to the oil companies"; and the State Department has been a favorite target of some politicians for supposedly harboring this sort of intention on behalf of U.S. business interests. But, in fact, American policy toward the conflict has only had passing reference to our role, or that of U.S. oil companies, in the area's energy production. Rather, our principal concern has been with Israel itself and with trying to reduce the opportunities in the area open to the Soviet Union because of the action-reaction nature of the Middle Eastern conflict: each round has only made the Arabs more dependent on the Soviet Union for arms and diplomatic support, and, despite their military withdrawal from Egypt last year, the Russians have duly taken advantage of the possibilities offered them. Meanwhile, increasing Soviet involvement helped Israel succeed in its requests for U.S. assistance.

When the Arab-Israeli conflict is discussed in the United States today, however, there is increasing concern about oil rather than about the Soviet Union. It is argued that at some future date the Arab oil-producing states could subject the United States to a form of blackmail because of its support for Israel. Until recently, this has not been much of a problem, because we have relied on the Middle East for only about 3 percent of our oil consumption. But would we not face a dilemma if we depended on Arab states for 25 percent or more?

The Problem of an Embargo

Like all such problems, this one turns on a combination of factors: opportunity, motive, effect and alternative. Could the Arabs enforce an oil embargo against the United States? Would they do so if they could? What would be the political consequences of any such action? And how can the problem be dealt with in advance? All these subjects are highly debatable, but the answers could be very important both for U.S. energy policy and for our overall policies in the Middle East.

To start with, the feasibility of an Arab oil embargo against the United States turns on the ability of the producer states to withhold enough production to make serious inroads into the total U.S. energy picture over a period of time, or at least to cause short-term disruptions that would do the United States some damage.

On the face of it, the higher the degree of U.S. dependence on Arab oil, the greater the economic effect an embargo would have. This would be especially true the more we depended on single countries for these resources. Thus, if Saudi Arabia does, indeed, reach a production level of 15 million to 20 million barrels/day by 1980, it could by itself have an effect on the U.S. energy position. This situation would be worse if the extreme estimates cited above for worldwide supply, demand and new investments are realized: by 1980 the entire world's reserve production capacity might be *less* than the oil production of any *one* major Middle Eastern Arab producer. And not only would a single government have great leverage, so the argument goes, but even civil turmoil (whether or not linked to the Arab-Israeli conflict) could be disruptive.

Even if this situation does not come to pass, there is still the prospect of cooperation by more than one producer. Indeed, an embargo by some combination of states, Iraq, Libya, Saudi Arabia, Qatar, and the other Persian Gulf states (members of the United Arab Emirates), could produce a very serious situation in the United States.

"If we stop selling them oil, maybe we'll be able to get across the street."

Padry in *Le Herisson*, Paris

How likely is that cooperation? Past experience does offer some guide. During the stoppage of Iranian production from 1950-53, the Arab states made up the shortfall, both to make more money and to secure better long-term markets. During the Arab embargo of 1967, Iran returned the favor; and it could be expected to do so again. But in a widespread embargo, non-Arab oil could not begin to meet U.S. needs projected for the 1980's unless there were a high degree of flexibility in energy marketing; that is, unless Iran and other non-Arab producers could shift their Europe- and Japan-bound production to the United States, while Arab suppliers met European and Japanese needs.

Of course, in 1967, there were strong pressures from the Arab producer states to lift the boycott against Britain and the United States as soon as possible; and it was lifted quite speedily, following an agreement of Arab leaders in which three oil producers—Libya, Kuwait and Saudi Arabia—elected to make sizable payments to Egypt and Jordan. This was a form of conscience money; it reflected the central preoccupation of intra-Arab politics with the conflict with Israel and the fact that the three oil states had not played much of a role in the Six-Day War (only Kuwait went so far as to send a few troops to Egypt).

In general, experience indicates that it is very hard for the producer states to hold the line on a political embargo beyond a brief demonstration of Arab solidarity. It would be especially difficult to hold it to the point of denying increased oil exports to Western Europe or Japan during an embargo directed only against the United States. Some oil—though how much is debatable—could then be diverted across the Atlantic and Pacific.

'They Can't Drink It'

This experience has a bearing on what is one of the most popular clichés about Arab oil: "they can't drink it." As the next chapter will point out, this cliché does not have the same force it once had; nonetheless, it cannot be entirely dismissed, especially in regard to countries interested in the benefits of long-term, stable markets.

This judgment also takes account of increasing cooperation among the Persian (Arabian) Gulf suppliers in bargaining with the oil companies on price, terms and equity participation in the production of oil. And it should be noted that there has been effective cooperation within the broader Organization of Petroleum Exporting Countries (OPEC), which includes Iran, Indonesia, Nigeria, Venezuela and the North African producers, as well as Arab producers in the Persian Gulf. Yet it is one thing for the Arab producers to cooperate among themselves (and with others) on *commercial* matters; it is quite another to extend that Arab cooperation into the political act of threatening the vital supplies of major powers, with the political or economic risks this would entail.

In a related area, serious calculations will also have to be made by both producer and consumer states about the threat of civil turmoil. For example, Palestinian groups might use blackmail—threats of sabotaging oil production or transport—to obtain financial and diplomatic support from the producer states. The producers might acquiesce or, just as likely, they might actually decrease their support for the Palestinian cause because they wish to maintain the long-term stability of energy markets.

Similarly, it is sometimes argued that U.S. energy supplies depend upon two lives: those of King Faisal of Saudi Arabia and Shah Mohammed Reza Pahlavi of Iran. Would a change in government in either country—or the "swallowing up" of Kuwait or a United Arab Emirate—lead to an embargo or reduced production for other reasons? This is always possible. Yet the economics of the matter would still be on the side of production (and might actually be strengthened under a government, say, in Saudi Arabia, that was more interested in the welfare of its people).

Possible Countermoves

The complex question of a possible embargo against the United States leads to consideration of some possible countermoves. For example, U.S. cooperation with Western Europe and Japan could be pursued, even if this would be of limited value. Here, timing is important. To establish consumer-state cooperation during a crisis might make matters worse, by bringing Arab wrath down on Europe and Japan. But if energy sharing agreements and practices were set up in advance as an aspect of cooperation designed to achieve many purposes, their implementation during an Arab oil embargo would be less politically upsetting; by their mere existence, in fact, they might reduce the possibilities of an embargo in the first place.

Of course, the European and Japanese oil situation is also becoming more precarious. In 1967, the United States had some spare production capacity—which is no longer the case—and Iran was able to play a significant role in offsetting Britain's loss of Arab oil—which it may not in the future because of the sheer magnitude of overall demand. Moreover, by 1980 Europe and Japan themselves may be importing as much as 20 million barrels/day from the Middle East, or nearly twice today's imports.

But do these forecasts mean that the United States must either restrict dependence on Arab oil producers or be prepared to modify its stand on Israel, perhaps abandoning it in a crisis? If the United States wished to cope with the problem in the terms defined here, it would have at least four other options: (a) diversify energy sources as much as possible, including the use of tax incentives to stimulate exploration outside the Middle East; (b) prepare standby rationing procedures; (c) stockpile oil; or (d) actually *increase* foreign imports, in order to provide for spare production capacity that would help us hedge against an oil cutoff of temporary duration. This last step would have high costs—including balance-of-payments costs—but it would reduce the potential impact of an embargo and might even make one less likely by minimizing its effects in advance.

How Important Is the Conflict?

Several political factors help to lighten the dark picture painted above, and indicate here, as elsewhere, the danger of assessing threats in terms of "capabilities" while ignoring the political dimension. One factor relates to the nature of the Arab-Israeli conflict itself in Arab politics and diplomacy. It is true that during a crisis the issue of Israel can overshadow almost anything else in many Arabs' minds. But this issue is still not all-consuming and, except in a major crisis, determines U.S. relations with Arab states far less than is often believed. Arabs, too, are capable of a pluralistic and many-sided view of the outside world. Much of U.S. concern about an oil embargo overlooks this fact. Thus, at the moment a decision to impose an embargo against the United States over Israel would require (a) evidence that the United States was directly helping Israel in an open conflict—as was alleged incorrectly during the Six Day War; and (b) a willingness of the producer states to jeopardize broader relations with the United States.

The same reasoning applies to an Arab embargo imposed in the absence of a crisis—i.e., in an effort to shift U.S. policy on Israel, generally. It would have to be evaluated in terms of how it would jeopardize other interests. Furthermore, much Arab opposition to U.S. policies in the Arab-Israeli conflict does not relate to the U.S. commitment to Israel as such, but rather to the widespread belief that the United States is not taking Arab political, economic and psychological interests seriously (including the economic needs of the Palestinian refugees) and has foresworn a more evenhanded policy on particular issues like the arms balance and the terms of a partial settlement.

Moreover, if there were an embargo, the United States would strongly resist being "blackmailed." We now clearly have a close relationship with Israel. Whether or not this is in our national interest—in strategic and economic terms—the sense of commitment is real and deep and is likely to continue. Even when we depend far more than today on Arab oil and natural gas, this feeling is apt to remain strong. In such circumstances, any effort to blackmail the United States over Israel would probably lift a crisis to the level of a serious confrontation with the offending states, going beyond economic conflict.

Assuming that all the preceding calculations prove grievously wrong, Israel itself is not helpless. If the United States stood aside in any coordinated Arab military attack, Israel could be expected to acquit itself very well. Of course, if the Soviet Union were involved on the side of the Arabs, Israel alone would not stand much chance. But then, the conflict would cease to be limited to questions of oil or Israel. The crisis would become a Soviet-American one, dwarfing the original cause.

Implicit here is the assumption that Israel will maintain sufficient strength to defend itself. But Israel should not suffer from a broader U.S. policy of helping to maintain an arms balance between combatants—even if the United States shifted its definition of an arms balance from today's *imbalance* in Israel's favor to one designed to afford deterrent protection for all the principal states in the conflict. And there is at least the possibility of Israel's again buying arms in Europe.

Finally, it is far from clear that the United States could force Israel to make any concessions, even if we tried to do so. Israel has never relied on U.S. support for its ultimate survival. It is most unlikely to do so in the future.

In short, an assessment of political possibilities in the Middle East indicates that increased U.S. dependence on Arab oil and natural gas is unlikely to affect the security of Israel significantly, even if we are less inclined to come to Israel's *diplomatic* assistance—on which it places far less value than being able to obtain military hardware from some outside source.

But a word of caution is in order. If this analysis is less pessimistic about the threat of an Arab oil boycott over the issue of Israel than is much commentary in the United States, the question still cannot be dismissed lightly. Certainly the U.S. perception of its freedom of action in the Middle East will not be *increased* when we depend more on Arab energy supplies.

Several possible U.S. efforts to reduce the risks even further have been mentioned. But these should not obscure the central point: that reducing the risks of a boycott is essentially a political problem, to be pursued through political means.

However, the United States itself may benefit from playing a less direct role in seeking a partial Arab-Israeli settlement. There is the continuing risk of disappointment, especially since any encompassing settlement is simply not in the cards—a fact that helps make even a partial settlement that much harder to bring about. Today, the United States is again regarded by several Arab countries as the key to diplomatic change, on the tenuous premise that we can "move" Israel. Disappointment of these hopes might only worsen a whole range of U.S. relations in the Arab world, and the process itself would continue to identify U.S. Middle Eastern policy with the conflict instead of permitting greater emphasis on broader relations with states in the area. At the very least, the United States should be more evenhanded in its approach to the Arab-Israeli conflict, either on the matter of arms or on support for particular Israeli or Arab diplomatic objectives.

Securing Cooperation

The role of the Soviet Union and the Arab-Israeli conflict have commanded most attention in past discussions of the energy problem. Yet in the future, there may be another development that could pose an even greater threat to the security of energy supplies coming from the Middle East. Quite simply, a number of producers may decide to restrict production for two reasons: first, to conserve a long-term source of revenue, rather than running huge inflows of a currency subject to devaluation and then seeing production slow down because of dwindling supplies. As noted earlier, Libya and Kuwait have already cut back on oil production, at least in part on conservation grounds, while Venezuela has some conservation worries.

Second, in a few years, some Middle Eastern producers with vast proved reserves will be earning so much money that—it is argued—they will be able to decide whether or not to maintain high levels of production. Indeed, they may elect to let oil appreciate in the ground rather than convert it into "money in the bank" that might not increase as much in value. Would the oil producers have to fear losing out to new sources of energy developed in the consumer countries? Perhaps in the long run; but for the foreseeable future, at least, the demand for oil may simply be too great. In the medium term, only if a price rise led to major increases in non-Middle Eastern oil reserves and production at economic cost, could the Middle Eastern producers find themselves losing out from restrictions on production.

This second possibility is the more serious one and applies to several producer states that the United States, Western Europe and Japan will rely upon most for oil. It is unlikely that Iran would be involved: the shah's complaints over the years have centered on his wish to expand production even beyond the desires of the oil companies, and he now projects that Iranian production will not level off until it reaches 8 million barrels/day. Nor would there be problems with Indonesia and Nigeria, both of which can use all the money they can earn from oil. But a country like Saudi Arabia, with oil production of perhaps 15 million to 20 million barrels/day by 1980, would have difficulty spending internally the amount of money it would earn even if it were disposed to do so. Depending on estimates made of price and production, Saudi Arabia could be earning net between $10 billion and $25 billion a year—an enormous sum.

What Uses for Oil Money?

Thus, increases in Middle Eastern oil production on which the Western world will depend may require finding significant uses for the vast pool of money involved. There are a number of possibilities. First, the oil producers could increase the amount of money they are spending on their own domestic development—although the amounts most could (or would) absorb are unlikely to meet the problem. Still, the United States, along with the other consumer states and the international development agencies, has a strong incentive to help these oil producers with something they often do lack—that is, technical assistance for development. They can also help to provide a general international atmosphere within which the development aspirations these oil producing states may have are taken seriously and supported.

Second, the oil-producing states could become deeply involved in economic development within the region. Kuwait, in particular, already makes some development funds available to other states. Abu Dhabi (a Trucial State) has a development bank on paper. Saudi Arabia has made some funds available to Somalia and the Sudan. The Arab Oil Congress, held in Algeria in 1972, raised the issue of revenue-sharing, though without result so far. And Egypt may work out some arrangements with Libya during the next decade to give Cairo greater access to the revenues from Libyan oil. However, the politics of common Arab development are difficult ones; and in view of traditional rivalries, the prospects for major Arab development efforts must be rated as only fair.

Third, revenues from oil could usefully be applied to economic development elsewhere in the world. They could be applied through multilateral institutions like the World Bank (about $500 million from the area has already been channeled in this direction during the past four years, although this is a trifling sum); or through bilateral arrangements with some individual countries. This would be a happy event: after all, the developing countries are being hit hardest, relative to income, by worldwide increases in the cost of energy. Unfortunately, this possibility has even less chance of coming to pass than does development cooperation within the Arab world. The idea of a two-tier price system (lower prices for poor countries) has not met with favor among the producers, in part because of fears that oil would be resold at a profit to the rich countries. Iran, however, has sold India some oil at a cut rate, partly as a way of demonstrating some independence of the Western marketing system.

Fourth, it is more likely that vast quantities of oil money will find their way into foreign investments of one kind or another. Indeed, the Saudi minister of oil and minerals, Ahmad Zaki Yamani, has already proposed direct investment in the United States in the downstream part of the oil industry. There are many other investment opportunities. Furthermore, investments in consumer countries would have value as "hostages," helping to insure that oil production continued.

In general, questions of guaranteeing energy supply need to be seen as part of evolving political and economic relations covering a wide spectrum of issues. If we want the producers to play the kind of role in energy that will be most beneficial to us, we will have to respond in terms that mean something to them.

Oil Money in the International Economy

A related problem is potentially as important as the implicit threat that one or more Middle Eastern and North African producers will voluntarily limit the production of oil. This is the matter of the vast quantities of foreign exchange that will be earned by the producers. Estimates vary, but total net earnings of these countries for investment could reach as much as $100 billion by 1980, and perhaps $150 billion by 1985. (These predictions depend both upon price and upon policies adopted by the consumers and may not be realized.) It is sometimes pointed out for purposes of dramatic impact that this latter figure is about the total amount of monetary

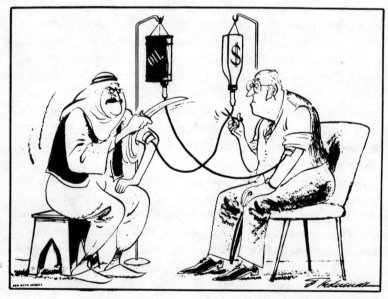

Behrendt in *Het Parool*, Amsterdam

reserves in the world today. But this comparison would be important only if major holdings were in liquid assets, which is most unlikely, beyond several billion dollars.

By any calculation, however, the oil producers will have large amounts of money for use in the international economic system. This means, quite simply, that the oil producing states will become major factors in the international monetary system, whether it continues to operate as it does today or is radically transformed.

Clearly, the major industrial centers will want these monetary reserves to contribute to stability in the international monetary system, rather than to cause disruptions. Some concern has already been expressed about the monetary crisis of February-March 1973, in which it appeared that some oil money increased the pressure on the dollar. By 1980, the pressure of oil money, as well as the pressure from funds controlled by the great multinational corporations, could be considerable. Nor can the holders of currency be blamed for trying to protect the value of their holdings against devaluation.

Several factors need to be taken into account. First, the consumer states can do whatever is possible to encourage the oil producers to invest their money in ways that would make it less liquid and, hence, less likely to be available to cause disruptions in monetary relations and exchange rates.

Second, concern about massive Arab monetary reserves is sometimes presented in terms of political threat: that these reserves might be used in a crisis over Israel to "bring down a currency." This possible Arab threat is seen as going beyond the shifting of reserves from one currency to another for purposes of protecting them against devaluation. And it ignores the tendency of Arab money managers—like their Western counterparts—to behave quite conservatively.

Even if a threat did materialize, a third point must be considered: Will the major industrial centers continue to subscribe to a system which either companies or nonindustrial countries will be able to affect in a serious way? To understand this point, it is important to understand the purpose of the monetary system in the first place. It is not autonomous; rather, it was created in order to make possible stable conditions for trade and capital flows. The monetary system is very much the tail on the dog; and if the tail comes to wag the dog to the point of causing serious discomfort—as may be happening now—then the system will be changed.

It is hard to imagine, for example, that a trading world doing $400 billion to $500 billion worth of business a year in 1980 will be subjected without complaint to changes in exchange rates brought about by movements of only several billion dollars in reserves, as can happen today. The industrial countries would also not accept a system that could be disrupted for political purposes, especially by nonindustrial states. Nor can one imagine the United States being content with a monetary system that would permit only several billion dollars of foreign-held reserves to cause important changes in an economy that by 1980 will have a GNP of $1,500 billion or more a year.

New Attitudes Needed

Nevertheless, because of the monetary reserves that will be controlled by the oil producing states, it is now essential to draw them more fully into the workings of the international trade and monetary systems. The objective should be to reinforce attitudes within oil-producing and consuming states—or to stimulate these attitudes where they do not now exist—that will foster common responsibility for the stable workings of the international economic systems.

This is largely a matter of psychology on the part of the producing and consuming states. If we want to reduce the risks that oil producers will withhold their precious commodity—for quite understandable economic reasons—then we must begin to recognize their importance. This means broadening monetary negotiations beyond the rich Western nations (the old Group of Ten); and even adding formal representation for oil producers in the new Group of 20 of the International Monetary Fund. Trade negotiations also need to be conducted with a much greater emphasis on the economic power of the oil producers. The key is in the attitude: if we can make

Projected revenues of the four major producing states of the Arabian peninsula

	$ billion	
	1973	1980
Saudi Arabia	4.70	11.0–25.6
Kuwait	1.90	3.1–6.4
Abu Dhabi	0.85	3.1–6.4
Qatar	0.35	1.9–2.9
Totals	7.80	19.1–41.3

The lower figure for 1980 represents the minimum projected production levels sold at the price scales laid down by the 1971 Teheran agreement. The higher figure represents the maximum projected production levels at a price tag (tax plus royalties) of $3.50 a barrel.

Projected state monetary reserves of the four major producing states of the Arabian peninsula

	$ billion	
	1973	1980
Saudi Arabia	5.00	30.0–75.0+
Kuwait	3.50	7.0–10.0+
Abu Dhabi	0.27	5.0–8.0+
Qatar	0.46	2.0–2.5+
Totals	9.23	44.0–95.5

The Economist Newspaper, London

the critical change, then the actual details are less important. But if we will not make the change—and fail to try engendering greater common responsibility for the international economic system on the part of the oil producers—then we may find ourselves with a very serious problem. In this context, trying to see the world as dominated by a global five-power balance of power, relegating other states to third-rate status, would only work against U.S. interests in promoting essential political relations with energy-producing states.

The industrial states also need to be more forthcoming on matters of trade preferences and other reductions of barriers to exports. Of course this is more true with regard to energy-producing states outside the Middle East, such as Venezuela, Nigeria and Indonesia, that are not at present largely limited to a single export.

The methods of bringing about a new role for the oil producers in international economic deliberations are less important than recognition in the West that it must be done. Even if changes made in the system by Western industrial powers could limit the potential threat of massive currency movements for political purposes, the threat of withholding production would still be present. Changing the monetary rules would help in a short-run crisis; but it is no substitute for a basic political approach. That political approach should be based on cooperation, and on acceptance in the rich countries—especially in the United States—of the fact that a few nations can no longer determine the rules of the global economic game all by themselves.

Cooperation in the West

In general it will also be important to foster cooperation among the Western industrial consumers of energy. The facts of Western European and Japanese dependence on Middle Eastern energy, along with increasing U.S. demand for foreign energy supplies, have already been presented. They indicate a real possibility of competition within the industrial world for Middle Eastern energy reserves. In limited ways, the producer states have already taken advantage of competition in order to increase prices and gain better concessional arrangements.

In theory, the consumer states have a strong incentive to cooperate with one another to try limiting the rise in prices for oil from the Middle East and North Africa. Whether they could have much effect, however, is a matter of widespread debate.

It is clearly important for the consuming nations to avoid possible conflicts with one another that could arise from the relations established by any of the three major industrial centers in the oil producing regions. For example, Sheikh Yamani's offer to the United States of a "special place" in Saudi energy, in exchange for preferential treatment for Saudi oil and investment in the United States, could set a poor precedent for U.S. cooperation with Western Europe and Japan unless due caution were exercised. The possible extension of the European Community into one or more producing states, through associate memberships, could cause concern in the United States and Japan. And from time to time Japan, Italy and France have made special deals with oil producers which have had an effect on other industrial nations.

Furthermore, the consumer states would benefit from adopting a number of possible joint steps, including rationing policies, energy sharing agreements, and perhaps even the expensive stand-by production capacity. They could also promote ventures to stimulate development of energy reserves outside the Middle East (as Japan, for example, is doing in places like Peru and even China).

However, it is impossible to separate the problem of energy from other problems of cooperation among these three industrial centers. In the United States, today, we tend to take for granted the friendly relations of the postwar period across the Atlantic and the Pacific. Yet there are now serious strains in our alliances with Europe and Japan: the international monetary system is still wobbly and in need of further reform; trade protectionism is growing; consultations on key political and economic issues fall far short of what is required. In the future, increasing demands for raw materials and energy, along with other factors, will place an even greater premium on amicable relations among the Western industrial powers, just as these developments will make good relations more difficult to achieve. This cooperation cannot center primarily on energy supply—an area in which common interest is only beginning to emerge and in which past efforts (in the Organization of Economic Cooperation and Development [OECD]) have borne little fruit. It must begin with today's other outstanding problems. Even then, cooperation on energy will be difficult to achieve.

In seeking cooperation, the Western industrial powers should also draw in the rest of the world's consuming nations. The price and availability of oil and natural gas, after all, are critical factors for virtually every economy, and especially for the countries of the developing world. As indicated above, for example, countries like Brazil or India may be hardest hit by any increase in the price of oil

61

by the OPEC states. India, for example—with 560 million people—would only need to have its imported energy bill go up by $2 per person a year to incur an added expense of more than $1 billion in foreign exchange. India simply does not have it and has already been forced by price rises to control some imports of oil. In like manner. worldwide increases in the price of oil are hitting all the poor consuming nations that are attempting to industrialize. The price in-

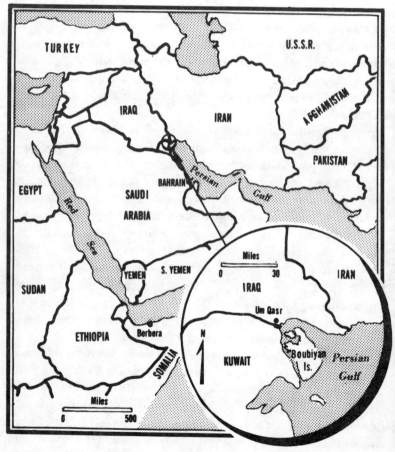

by Joseph F. Mastrangelo—*The Washington Post*

creases of the past two years may have increased the total import bill of poor countries by 2 percent or more, i.e. about $1 billion a year.

Bringing these other nations into any bargaining process is thus a matter of equity—a realization that otherwise development will be set back in many countries and the gap between rich and poor widened even further.

Bargaining on Price

Clearly it is in the interest of the United States, along with all other consuming nations, for the world price of fossil fuels to be as low as possible. What can be done to secure this goal is another question. Some experts contend that effective consumer-state cooperation would do an effective job of stopping price rises on the part of OPEC nations, even if it failed to roll prices back.

Some observers also contend that bargaining would be more effective if governments were directly involved, instead of leaving the bargaining to the companies. According to this line of reasoning, individual producer states could be "picked off" one by one in their scramble to increase shares in the total energy market.

Other experts challenge this line of reasoning. They argue that the consumer states are more likely to be picked off, especially in a world of oil shortage and in view of the possibility that one or more oil states could limit production without suffering long-term economic losses. So far, OPEC has also been more effective than the OECD nations in holding the line on energy bargaining, although the situation might change if there were a true "organization of petroleum *importing* countries."

A major argument has developed about the role that the producer companies play in negotiations on price. This is the "Adelman crisis"—as spirited as the energy crisis itself—named after Professor M. A. Adelman of the Massachusetts Institute of Technology. He contends that the oil companies can only gain from price increases—by passing on costs in a high-demand economic situation—and, therefore, do not bargain very hard. Other experts disagree with him and argue further that the consumer countries could not possibly replace the role of the companies in negotiations.

Avoiding Confrontation with Producer States

Whichever school of thought is right on these issues, it is all too likely that any bargaining with producer states, whether done cooperatively by consumer states or not, could lead to a serious confrontation with producers and to a general climate of acrimony. Such a development will not be to anyone's advantage, especially in view of the principal interests on each side: among other things, the producers will want stable markets and sources of technical assistance to exploit their energy; the consumers will want stable sources of supply and prices that are under control. The consumer states, therefore, have a strong incentive to work for the most conservative behavior on the part of producing states, even if this means both accepting somewhat higher prices than might otherwise be possible and not bitterly opposing a progressive divestment of ownership and control.

This divestment has already begun: in the Persian Gulf, the producer states are now purchasing a 25 percent interest in oil company production—"participation"—and will extend this control to 51 percent in 1982 under present agreements. Meanwhile, Iran will take over control of its production in 1979, and Libya, Algeria and Iraq have been following the more traditional route of nationalization. The consumer states should not suffer from these developments, however, and some companies will still have a major role to play in all phases of oil production and marketing.

Cooperation with the Soviet Union

The issue of cooperation among consumer states—and of avoiding a serious clash with the producers—inevitably raises again the role of the Soviet Union and the possibility that it (along with the East European states and, one day, possibly even China) could become a conservative influence in energy trade. Furthermore, in the event that the non-Communist consumer states elect to confront the producers, particularly over the security of supply, the question of a military sanction may be raised.

In today's world, however, there are real limits on the use of gunboat diplomacy, just as there are limits on the military activities of the major powers against one another. Yet would the United States, or some other consumer states, be tempted to use military power in the Middle East if it felt this was the only way to secure supplies? And would the Soviet Union stand by? This possibility has already been raised by some prominent Arab observers in the Middle East, as an argument for caution. Nor must it be discounted out of hand.

However, it is very doubtful that even a degree of Soviet-Western cooperation in energy would lead either side to accept the use of military force by the other, short of some truly vital threat that would somehow involve them both. To do so would require a higher degree of mutual trust than is currently possible in East-West relations. Neither side could be sure that the other's military action against a producer state would not lead to larger ambitions.

If a military sanction against a threat to truly vital consumer interests nonetheless remains implicit, it is important for all powers concerned with energy and the Middle East to scotch any idea that military power might become the arbiter of disagreement. Otherwise, relations would be soured from the outset; cooperation might become impossible; and all the consuming nations might face far greater problems of energy supply than they could hope to resolve by resorting to an anachronistic use of arms. Here, too, there is no real alternative to making the political efforts that alone can lead to productive relations in energy supply.

Paying the Price

The United States will be paying an increasingly high cost for imported energy. It was $2.2 billion (net) in 1965, $3 billion in 1970. From now on (the estimated 1973 cost is $6 billion to $7 billion) the curve will rise steeply. Some experts have predicted that the United States will spend up to $30 billion in 1985 to import energy. This is certainly a dramatic figure: about two-thirds of total U.S. imports at the present time.

Projections like this one have led many observers to argue that the United States must speed up use of domestic sources of energy, even at higher cost, in order to protect the U.S. balance of trade, already heavily in deficit.

How sound is this view? A number of observations are in order. To begin with, growth of U.S. energy imports implies that the U.S. economy itself is growing. If so, then our exports will be going up, as well. How much is highly speculative. But at least the raw figure of a $30 billion yearly import bill needs to be viewed in context.

Second, however, even if our exports do go up by some amount as a simple function of economic growth (as opposed to a further devaluation), they are unlikely to go up at the same rate as our energy imports. This is so because of the shift in the percentage of our energy that will come from abroad. Oil imports currently account for 33 percent of consumption. By the end of the decade they will account for 51 percent or more if major efforts are not made to limit demand or accept the far higher costs of domestic alternatives available between now and then.

Financing Imports

As a result, the United States will have to find some way to finance its added imports. Of course devaluation is always an alternative to running chronic deficits in the balance of payments. At some point this might be necessary. There are four objections, however: (1) the producer states now demand payment in constant dollars, and have sought price increases after both U.S. devaluations; (2) foreign holders of U.S. dollars might object to the dollar's losing even more value and insist that we meet our domestic economic problems instead; (3) they might disapprove of unfair competition from the U.S. in exporting our energy-payment problem, especially if they have a similar one; and (4) devaluation is no substitute for long-term adjustments in the domestic economy. Used as a substitute, it does make us poorer.

Thus, whatever the other solutions to the problem of paying for energy imports, we have to make our exports far more competitive in world markets than they are today. Nor is this problem limited to energy supply; it is critical for our entire trading position in the world.

For an aging economy like that of the United States, becoming more competitive in world markets is not as simple as it sounds. Indeed, there are strong pressures today to increase trade protectionism as a way of avoiding the whole problem of competitiveness—although protectionism would decrease the real income of all American consumers and would slow down economic growth. Some economists also argue that it will be difficult for the United States to become more competitive, simply because the emphasis in U.S. production is shifting from goods to services, most of which cannot be exported in any event.

67

Nonetheless, there is good reason to shift American production into areas where we can compete more effectively with foreign producers. Agricultural exports are a case in point: exports of grain and soybeans will be critical in paying for energy imports. Furthermore, the U.S. government should undertake a major program of adjustment assistance—perhaps $500 million to $1 billion a year—to help workers and capital move to more productive industries. This is an economic necessity; it is important on social grounds, so that individual workers will not suffer the impact of our goods being priced out of world markets; and it will be necessary as one measure to reduce the opposition of major elements of American labor to liberal trade that promotes our high standard of living. Even then labor's opposition will be very real indeed, especially because the adjustment assistance program under the Trade Expansion Act of 1962 worked so abysmally.

Paying for Greater Self-Sufficiency

Despite all the qualifications advanced in this discussion about an energy crisis, we may well conclude that we should maximize our self-sufficiency in energy. This does not mean that we could eliminate some growth in energy imports from abroad, at least not for many years. But it does mean that some measures can be taken toward this end.

To be sure, investing heavily in domestic energy will be important for the long term, simply because of the limits on the world's fossil fuels, even if new discoveries or new processes considerably extend the size of what is believed will prove to be total reserves. But we should weigh very carefully the costs of developing domestic reserves. In terms of investment, for example, developing U.S. fossil fuel reserves—or other forms of domestic energy—will be very expensive. In addition, if the United States limits its imports of energy (e.g. through a tariff or an extension of quotas), and accepts higher domestic prices, our balance of trade would be affected, because the cost of U.S. goods would tend to go up relative to those of other countries. The price rise might not be significant if the energy component of finished goods proved to be low or if other countries also faced rises in the price of energy and hence of their export goods. Nevertheless, the general problem of competitiveness in goods would remain and—without a further devaluation—would give us a large trade deficit even if we imported as little oil and natural gas as possible.

One major qualification must be added, however: if Middle East prices continue to rise, the differential cost of U.S. domestic energy versus imports might disappear and thus help stimulate domestic production without a balance-of-payments penalty. However, there would still be a limit on how fast the United States could tap other reserves or other fuels.

There would also be a price tag attached to any effort to reduce foreign energy dependence for national security reasons beyond efforts to develop new reserves and methods of using existing ones. Stockpiling is the most obvious example. Perhaps a stockpiling program would be advisable because of the uncertainties that are indicated in the preceding analysis. But it would be at a cost which would not even provide the long-term benefits that could be derived from high capital investments in developing nuclear, solar or coal-derived energy. At levels of consumption for 1970, it has been estimated that a stockpiling program good for 90 days' independence of Middle Eastern suppliers would cost the OECD nations over $600 million a year. At 1980 consumption levels the figure would be far higher, depending on the number of days' stocks held. With consumer-state cooperation, this might only have to be a short period. After all, a threat to the energy supplies of all three major industrial areas at once would pose a very serious political threat that would raise the level of confrontation far beyond the immediate issues of energy supply. What is involved in stockpiling is a hedge against short-run disruptions—which might result from a tanker shortage, a technical mishap or sabotage to production facilities in the Middle East.

In addition to stockpiling, the United States could develop a standby rationing program. Still another alternative, as suggested earlier, would be to create a standby production capability that could take up the slack during any short-term disruption of some imports. This would clearly be most costly if it were added to existing production, and would have to be limited for all the reasons advanced earlier about U.S. reserves and production. The idea, however, would be to increase our foreign dependence on energy, even reducing our domestic production, while retaining the ability to resume higher levels of domestic production very quickly. The economic trade-off is between (a) investment in standby capacity and a higher import bill; and (b) saving that investment and possibly having a stockpiling program. The political trade-off is between two very important concepts of security: the conventional school warns of dependence on foreign imports; another school warns of a depletion of U.S. fossil fuel reserves. Far better, this latter school argues, to deplete foreign reserves and conserve our own for the future. Indeed, this idea partially explains our acceptance of oil imports in the past.

In sum, as in all calculations of security requirements, the cost factor is important in judging which threats are worth insuring against and which are not. Threats to truly vital interests, of course, must be met whatever the cost. But in the case of less-than-vital interests—with which, despite present concern, maintaining self-sufficiency in energy supply must be included—comparisons between costs and benefits increase in importance. Furthermore, since the American consumer is significantly affected by the price of energy, these questions should be given widespread public debate.

The Environment

One final set of costs must also be considered: the environmental. We have already discussed those involved in limiting energy demand and expanding domestic supplies. But we must also recognize that there are environmental costs involved in importing more energy as well, or getting more oil offshore or from areas where it must be shipped. The foundering of the Torrey Canyon and the oil spills off Santa Barbara offer object lessons: importing more oil or

Oil pipeline stored in Fairbanks, Alaska   An Exxon photo

producing it offshore increases the chances of ocean pollution. We no longer regard the oceans as bottomless sinks that can absorb all the pollution we pour into them. Consequently, we must face questions of greater regulation of ocean transport and higher costs in guarding against spillage.

Importing energy into the United States means more ports—especially deepwater ports—to pipe oil and LNG ashore. The risks and costs here, too, must be considered, and there is already rising public opposition in many parts of the United States to the building of the needed facilities. Yet here the trade-off with domestic environmental damage may tip the argument in favor of imports. The same is true of an increasing tendency to import oil that is already refined, rather than to tolerate more domestic refineries. In effect, we have so far chosen to "export" pollution to other countries, many of which do not have our pollution worries or are still at the stage where economic gain is prized more than environmental quality.

In a very basic sense, the issue of the environment is also one of security. In considering overall energy policies, we need to consider all the different kinds of security—the political, economic and environmental—involved in increasing energy demands and changing sources of supply. At the moment, the environmental question is one for the United States and other industrial nations to worry about. In time, however, there could be a broader issue of worldwide tolerance of pollution in the use of energy. Exporting pollution does not stop it; it merely shifts its point of origin. We may face important questions of equity in the use of energy, not unlike questions posed about the division of the world's limited fossil fuel reserves, both among industrial countries and between rich and poor. At the very least, the rich countries should bear the major burden of developing and using forms of energy that pose less of an environmental threat.

Conclusion

Is there an energy crisis? This question may now be easier to answer than it was at the start of this discussion. In the United States, we will face a rise in the price of the energy we use, either now or later; we will face increasing threats of pollution from the production and use of energy; and whatever we do today, at least for some years, we will have to import more energy from abroad. None of these spells crisis, however, provided that we act intelligently on what we know and are careful to make plans that hedge against the uncertain future.

The elements of such intelligent action and planning as they have emerged from this discussion can be summarized thus:

Domestically we should (a) elaborate a comprehensive national energy policy; (b) develop more accurate statistics on both the supply and demand of energy; (c) focus far more than we are presently doing on limiting our demand for energy; (d) increase investment on a major scale in diversified research, development and production of existing sources of fossil fuels and newer forms of energy; (e) intensify efforts to find and exploit domestic sources of fossil fuels; (f) conduct a major debate on the impact of energy demand and supply on the environment; and (g) undertake a vigorous program of adjustment assistance.

In our foreign energy policy we should (a) consider seriously in public debate the various cost factors of energy imports; (b) diversify our sources of foreign supply; (c) adopt a standby rationing program and investigate the advisability of a modest stockpiling and reserve capacity program; (d) improve political relations with Canada and Venezuela; (e) import some energy from the Soviet Union, while insuring that any dependence is mutual and pursuing détente in the Middle East; (f) continue to support the search for a partial settlement of the Arab-Israeli conflict, being more cautious and evenhanded in our involvement, while promoting better relations with Arab states; (g) do what we can to promote the use of oil money by Middle East producer states in economic development and investment both inside and outside the region; (h) involve the major oil producing nations in the workings of the international monetary and trade systems; (i) cooperate with Western Europe and Japan in energy-sharing agreements and a cautious approach to price-bargaining; and (j) involve the developing countries as consumers of energy in efforts to insure a stable supply of oil and limit the rise in price.

More than anything else, however, we can meet our concerns about the supply of energy only by changing some basic attitudes about ourselves and about the world we live in—not out of altruism or whim, but because we must.

In the process, we must realize that an energy crisis is not an isolated event in our world. Energy, after all, is only one area of our nation's economic life and only one area in which we are finding ourselves far more involved in the outside world than before. In the future, we can also expect a much greater dependence on other countries for a wide range of other raw materials which either have no substitutes at home or where substitution imposes high costs. Moreover, the relatively faster growth of Western Europe and Japan is posing major economic difficulties for us, although foreign trade, itself, continues to make up only a small part of our GNP. Growing competition in trade, shifts of power within our traditional alliances, the emergence of the multinational corporation, and a wide range of international monetary and commercial problems promise us no respite. We are becoming a permeable society with no real alternative to interdependence with other nations.

Graham in
Arkansas Gazette

Thus, we could struggle to become more self-sufficient in energy and find that our other forced involvements in the outside world would still make us vulnerable. But we cannot have a Fortress America economically any more than we could have it militarily. This is not a palatable thought for most Americans, nurtured on the Farewell Address of President George Washington and inheriting a legacy of national power and self-sufficiency. Indeed, the next few years will be among the most difficult in our history in relating to other countries and peoples—as difficult, perhaps, as our entry into world politics only three decades ago.

By all means we should take seriously the supply and price of energy; but we must keep them in perspective as well. Any lasting solution to what is seen to be an energy problem must be sought not primarily in the wonders of technology but rather in the realm of politics, including the domestic politics of conservation. More specifically, it should be sought within the broader context of our evolving relations with other countries; within new institutions of international cooperation and old institutions that are revised and updated; in a renewal of our search for an international economic order in which many nations have a stake; in diplomacy and economic adjustment rather than in confrontation politics and military power; and, finally, through seeing ourselves as part of the outside world, and not as somehow separate. Only in these ways can we develop a vision of the world and of the future that is sufficiently broad and deep for us to begin coping with the difficult era ahead.

AFTERWORD

On April 18, as this *HEADLINE SERIES* was going to press, President Nixon sent his long-awaited energy message to Congress. From the standpoint of this analysis, it contained three important and forward-looking statements: (a) the President spoke of an energy *challenge* rather than a crisis; (b) he urged adoption of a "national energy conservation ethic"—though with few specific details beyond the progressive decontrolling of natural gas prices; and (c) he directed attention to all types of energy, and not just to oil and gas, which are the subject of concern because of increasing imports.

Most significantly, the President abolished the Mandatory Oil Import Program (quotas); replaced the 10.5 cent tariff on each barrel with a license fee (equivalent to the tariff) on imports and qualified the fee system to permit increased imports of inexpensive oil for the short term; urged increased Federal expenditures on research and development of domestic energy supplies (but, at $771 million for fiscal year 1974, fell far short of the $2 billion a year advocated by Senator Jackson); and urged greater exploration for domestic fossil fuels and relaxation of environmental standards on the use of coal.

However, the energy message said virtually nothing about the foreign policy problems of energy dependence. Nor did it provide any arguments for the goal of increasing self-sufficiency in energy upon which the message was premised. The President did stress problems of the U.S. balance of payments; and he referred several times to an undefined concept of "national security." Without elaborating, he endorsed the pursuit of "national interests through mutual cooperation [rather] than through destructive competition or dangerous confrontation." And he specifically endorsed oil sharing with other OECD countries in times of acute shortage; cooperation on research and finding "ways to prevent serious shortages," including "international mechanisms for dealing with energy questions in times of critical shortages"; and study of incentives to provide domestic storage and excess production capacity.

Unfortunately, despite the President's projection of a short-term increase in oil imports, his message cannot be called "international." It does not relate U.S. energy problems to those of other countries in any significant way; nor does it face the need for Americans to accept greater involvement in the outside world. Again, we are led to believe that our own domestic efforts— primarily in supply rather than demand— will in time permit us to avoid increasing dependence on other countries. In sum, this message does not provide the basis for national debate about the foreign policy and foreign economic problems of energy. And it does not begin to adapt U.S. thinking to critical changes in our involvement in the world—changes which will inevitably shape U.S. foreign policy and the way we live during the 1970's.

Talking It Over

In this discussion guide you will find discussion questions and reading references. These are suggestions only—a starting point to help you plan a study-group program or a classroom teaching unit.

Discussion Questions

How and why is U.S. dependence on oil imports changing?

What are some of the alternatives to oil as a source of energy? Do you think they offer the prospect of significantly relieving our energy needs over the next decade and a half?

What are the major problems involved in obtaining increased oil imports within the Western Hemisphere?

What are some of the strategic implications of greater dependence on the Soviet Union for increased supply? Under what conditions, if any, would you favor greater reliance on the U.S.S.R. for energy imports?

How and why does the Arab-Israeli conflict affect our energy situation? In view of our increasing dependence on Middle East oil, do you think our present policy toward the Arab-Israeli conflict is sound? Why or why not? If you favor altering our policy, what changes would you recommend?

Do you think there is any realistic possibility that the Arab states might impose an embargo on shipments of oil to the United States? If so, what measures do you think the United States should take to forestall such a contingency?

What are the major economic problems that will arise from the inflow of dollars and other convertible currencies into the oil producing states of the Middle East in payment for oil? What measures should we and the other major oil importing nations take to deal with these problems? What problems does the United States face in financing oil imports? What measures should we take to deal with them?

How might the oil producing states use their energy revenues to strengthen the international economic system?

To what extent is the energy "crisis" a function of the American consumer ethic? Do you think that changes in the consumer ethic in the direction of reducing energy demand offer a feasible way of dealing with the crisis? Would you, for example, favor rationing of gasoline for automobile pleasure driving? putting greater reliance on the use of mass transit to conserve energy?

What are the implications of the energy problem for the interdependence of nations? How does the interdependence of nations bear upon possible solutions to the energy problem?

What are some of the environmental costs of increasing energy demand? Do you think we shall have to lower standards of environmental protection to meet our energy needs? If forced to choose, would you favor lowering standards or decreasing consumption?

Do you think that President Nixon's proposals are adequate to deal with our energy problem? If not, what solutions to the problem would you offer?

READING REFERENCES

Adelman, M.A., "Is the Oil Shortage Real?" *Foreign Policy,* Winter 1972-73.

Akins, James E., "The Oil Crisis: This Time the Wolf is Here." *Foreign Affairs,* April 1973.

Campbell, John C. and Caruso, Helen, *The West and the Middle East,* New York, Council on Foreign Relations, 1972.

Conant, Melvin A., "Oil: Co-operation or Conflict." *Survival,* January-February 1973.

Darmstadter, Joel, *International Flows of Energy Sources.* Washington, D.C., Resources for the Future, July 1970.

_____ and Hunter, Robert E., "Energy in Crisis?" in *The United States and the Developing World: Agenda for Action.* Washington, D.C., Overseas Development Council, February 1972.

"Foreign Policy Implications of the Energy Crisis," Hearings before the Subcommittee on Foreign Economic Policy of the Committee on Foreign Affairs, House of Representatives, 92nd Cong., 2nd.sess. Washington, D.C., USGPO, September 1972.

Hunter, Robert E., "In the Middle in the Middle East." *Foreign Policy,* Winter 1971-72.

_____, *The Soviet Dilemma in the Middle East, Part II: Oil and the Persian Gulf,* Adelphi Papers No.60. London, The Institute for Strategic Studies, October 1969.

Levy, Walter J., "An Atlantic-Japanese Energy Policy." European-America Conference, Amsterdam, March 27, 1973.

Lewis, Richard S. and Spinrad, Bernard I., eds., *The Energy Crisis.* Chicago, Ill., Educational Foundation for Nuclear Science, 1972.

Mansfield, Peter, ed., *The Middle East: A Political and Economic Survey,* 4th ed. New York, Oxford University Press, 1973.

Nixon, Richard M., *Message to Congress on Energy,* April 18, 1973.

The Growing Demand for Energy. Rand Corporation Interview Report, April 1971.

The Oil Import Question: A Report on the Relationship of Oil Imports to the National Security, Cabinet Task Force on Oil Import Control. Washington, D.C., USGPO, February 1970.

The Potential for Energy Conservation: A Staff Study. Office of Emergency Preparedness, Executive Office of the President. Washington, D.C., USGPO, October 1, 1972.

United States Energy: A Summary Review. U.S. Department of the Interior, Washington, D.C., USGPO, January 1972.

U.S. Energy Outlook, An Initial Appraisal, 1971-1985. Washington, D.C., National Petroleum Council, July 1971.

U.S. Energy Outlook, A Summary Report. Washington, D.C., National Petroleum Council, December 1972.

"U.S. Interests in and Policy Toward the Persian Gulf," Hearings before the Subcommittee on the Near East, Committee on Foreign Affairs, House of Representatives, 92nd Cong., 2nd.sess. Washington, D.C., USGPO, February, June, August 1972.

World Energy Supplies 1961-1970, Statistical Papers, Series J, No.15. New York, United Nations, 1972.

Standard References: *The Arab World Weekly, BP Statistical Review of the World Oil Industry, Oil and Gas Journal, Pacific Basin Reports, Petroleum Press Service.*

94

FOR A VALUABLE RESOURCE FOR YOUR CLASSES

and a way to keep up-to-date on key foreign policy topics in the news...

Subscribe to **HEADLINE SERIES,** published five times a year.

Each issue
- is about a major world area or issue
- is written by a noted scholar
- is brief (usually 64 pages)
- is highly readable
- includes basic background, maps, charts, discussion guides and suggested reading

- -

SUBSCRIPTION ORDER FORM

Foreign Policy Association
345 East 46 Street, New York, N.Y. **10017**

Date_____

Please enter my subscription to HEADLINE SERIES

☐ 1 year $6.00   ☐ 2 years $11.00   ☐ 3 years $15.00

Name _____

Street _____

City _____ State _____ Zip_____

☐ Payment enclosed   ☐ Bill me

- -

Titles of past issues on topics of current interest

215	The Soviet Union & Eastern Europe, New Paths, Old Ruts	by Robert G. Kaiser and Dan Morgan	1973
214	Understanding India	by Phillips Talbot	1973
213	Dollars, Jobs, Trade & Aid	by William Diebold, Jr.	1972
212	The Interdependence of Nations	by Lester R. Brown	1972
211	Latin America Toward a New Nationalism	by Ben S. Stephansky	1972
210	Black Africa: The Growing Pains of Independence	by L. Gray Cowan	1972
208	Rethinking Economic Development	by Robert d'A. Shaw	1971
207	Cuba and the U.S.—The Tangled Relationship	by Robert D. Crassweller	1971

Price per copy: $1.25

Orders for single copies must be accompanied by payment if they amount to $3.00 or less.

Quantity Discounts

10-99 .... 25% off   500-999 .... 35% off
100-499 .. 30% off   Over 1000... 40% off

For information on all currently available HEADLINE SERIES and other FPA publications, write to Foreign Policy Association for latest catalog.